SLIDERS:
The Enigma of Streetlight Interference

Hilary Evans

Anomalist Books
San Antonio * New York

An Original Publication of ANOMALIST BOOKS

SLIDERS: The Enigma of Streetlight Interference

ISBN: 1933665475

Cover image: Crystal Hollis

Book design by Seale Studios

For information, go to anomalistbooks.com, or write to:
Anomalist Books, 5150 Broadway #108, San Antonio, TX 78209

Contents

Dedicated to the SLIders from many countries
who told us what happened to them.

Before you begin

Streetlight Interference (SLI) is an alleged phenomenon, in which it is claimed that certain people, passing near a streetlight at night, cause it to spontaneously extinguish (or if off, come on). Although there are hundreds of reports by both SLIders and witnesses, the subject remains controversial.

Though trivial at first glance, closer study shows SLI to be a complex process, rich in paradoxes and contradictions. If true, however, the claims carry profound and exciting implications for science and for our knowledge of human potential. This book reviews the evidence and seeks a scientific explanation. Every account is taken from SLIders' first-hand reports of their experiences.

Abbreviations

SLI – Streetlight Interference: this generally involves causing a streetlight to go off if it's on, on if it's off.

SLIDE – the Streetlight Interference Data Exchange, which acts as a receiving center for SLI reports and comment, in association with ASSAP (see below).

SLIders – people who ostensibly produce SLI effects

SLIABILITY – the ostensible ability of certain individuals to unconsciously cause SLI-events.

SLI-force – the force that SLIders seemingly use to interfere with streetlights and other appliances. This may take the form of a Force Field.

SLI-state – the condition of mind or emotional set of the SLIder when doing SLI.

SLI-books – books about SLI: so far there is only one – the one you are holding – but hopefully there will be more.

ASSAP – the Association for the Scientific Study of Anomalous Phenomena, based in London, England, which published the first report on SLI in 1993 and continues to investigate the phenomenon. On the internet at www.assap.org.

Other notes

The number [xx] after a citation refers to the individual informant who contacted us; these are listed in an appendix at the back of the book. To protect their privacy, complete names and addresses are not supplied, but their original letters and emails are on file with ASSAP. Their occupations are given, if known.

Acknowledgements

Thanks to all those who have contributed accounts of their experiences, and/or made comments and suggestions, Dr. Richard Wiseman, and especially Mary Rose Barrington for her account of JOTT. A list of SLIders who sent us narratives and views is included at the back of the book.

Chapter One

What happens when SLI happens

> Dear Hilary, What an absolutely ridiculous and outrageous idea! – except that it has been happening to me for several years, and I thought I was the only one! – a resident of Arroyo Grande, California. [103]

People respond to Streetlight Interference with amusement or amazement, with belief, disbelief, or relief at learning that it happens to others as well as to themselves. SLI takes many different forms, and people react to it in many different ways.

The best way to discover what happens when SLI happens is to have it happen to you. For those of us who have not had the benefit of personal experience, though, the next best way is to learn what SLIders say happened to them. A SLIder in Surrey, England, reports:

> When living in London I was able to do it almost at will – I have as witnesses my husband and a large group of friends who, over the years, came to regard my ability as an endearing trait. [118]

A ridiculous idea? An endearing trait? SLI experiencers react in every sort of way when it happens to them. A few more accounts:

> *A self-employed salesman from Illinois:* The most recent event occurred just before Christmas, near my home in a shopping mall. There were a

great many lights in the lot: it was snowing that evening. One went out as we got out of the car. My children and I exclaimed, "We got one" as we usually do. On our way out the same light went out again. We started driving around the lot looking for another "victim." We would pick out the ones that looked a little dimmer and drive under them. If it did not go out right away we would wait or yell POW (just for effect I think) until we were able to knock out around six, including the same one three times.

Another incident occurred in Indianapolis. I just happened to be discussing this phenomenon with a couple of business friends over a drink. One of them had witnessed this on a prior occasion. The other of course thought we were nuts. When we went to the parking lot to get to our car, I got one. We all laughed at the "coincidence." As we were driving to a restaurant and waiting at a stop light, another went out. More laughter from us as we just converted another non-believer! [163]

Not many SLIders get to where they can play games with their ability, but this account brings out the take-if-for-granted attitude that long-term SLIders frequently adopt. Few SLIders are worried or disturbed even when they come to realize that they differ from most folk in doing what they do.

A young man from Harrogate, Yorkshire, England: My experience of SLI started in 1991 when I was 19. I was walking back from my girlfriend's house after a night out followed by a smooching session at her house. It was summer-time, in the small hours of the morning, and quite cold. I was walking through an open park lit by old-fashioned, ornate streetlights, when the nearest light-post to the footpath went out as I approached. I thought nothing of it except "the bulb has gone." The very same thing happened the following night, and again I dismissed it as a dodgy bulb and thought nothing of it until it happened the next time I passed, at which point I started to think something was up.

Over the months as I returned home from my girlfriend's house, the light would always do the opposite as to its original state, i.e. if it was off it would turn on and vice versa. After passing the light-post it would usually revert to its original state – I would usually be over 100m from it when it did this. On the approach it would vary from 60m to 10m past the post before it would turn on or off.

I told my friends about the situation and to silence the skeptics and piss-takers I took them past the light-post one dark evening. Sods' law! as nothing happened, not even a flicker.

It wasn't until I noticed I had the same effect on other light-posts that I noticed the conditions for the "light-post thing" to happen. That is, it was usually late at night (after 9 pm, say), dark, quite chilly, and my state of mind. Usually I would be quite tired, on edge, nervous of my surroundings (fear of mugging, I suppose) and I reckon my adrenalin levels will have been up. This sort of explains why I couldn't "perform" in front of my friends, having been in a relaxed situation. I have since shut my friends up as I have shown my ability on more than one occasion. I can only think that the cause is over-active brain waves, a result of my nervous emotions... [178]

This account is typical in the way a SLIder gradually comes to realize that something out of the ordinary is going on. Also characteristic is that it slowly dawns on him that his state of mind is a crucial factor.

> *A professional magician from London, England*: The moment I saw the header of the article in *The Independent* I knew exactly what it was about as I have experienced this on a number of occasions. The first time I realized something was happening was when I was about 17-18. (I am now 32) I was taking my dog for a walk. I noticed that lights were going out when we walked under them and then flickering back on when we had passed. It was only noticeable because of the number involved. It didn't frighten me but I became conscious of it. I remember walking under them trying to make them go out but I couldn't. The moment I stopped willing it to happen it would start again – like someone catching me out. I sort of anticipated it for a while and didn't really tell anyone about it. A few years ago I noticed it happening again – the first time for a long time. Again, I was with my dog and this time we turned out a number of lights in a car park across the road. I told a close friend when I got home and he came out to watch from the other side of the road. As we walked around the park they all went out as we passed under them, and then came back on when we had moved away. I was so chuffed that it had happened while he was watching. I seem to recall that both periods coincided with stress, some of it quite intense. [132]

Later he wrote to tell us that he had just been on holiday to New

Zealand with his girlfriend and turned out lights in Auckland. As a professional magician he takes a special interest in the phenomenon and speculates as to whether it could be anything like the feats associated with Israeli wonderworker Uri Geller, but he still finds them puzzling.

> *An office worker from Texas:* The reason I am so very interested is because I myself have a lot of various phenomena around me. Streetlights diffuse and go off at night, sometimes very slowly, sometimes very quickly. Streetlights go on in the full sunshine at times, when I come close to them. I have had periods in my life when several appliances broke down, just one after the other. Regular light bulbs will also come on or go off when I am around, either in people's homes, or in restaurants – same thing with fluorescent lights. I have had electronic garage doors open on me twice, once at my house in Lexington, Kentucky, another time in my home in Ocala, Florida. The garage doors would open about halfway and then go up and down rather quickly in a crazy way until the energy dissipated and ran out.
>
> If I am on really high energy, which I am about half the time, I can hold a compass and the needle will go crazy and just go around and around, until I let go of the compass. I also periodically have a tremendous amount of trouble with my handheld tape recorder. It will stop recording, but is still capable to play back. I went through about 10 of them over a period of a couple of months. Once it was so bad it even wiped out what was on the tape. At other times several of the fuses and lights in my car will blow. Once I was really excited. I was in my travel trailer at the time and

> when I reached out to turn on a light, there was like blue lightning running through the little bulbs (two of them) and a fuse blew, and all the DC lights in the trailer went out.
>
> I also can get very charged with static electricity, so much so that sparks actually fly around me and if anyone else is close by, the sparks will connect with them. I used to have the same sort of problems with an IBM copying machine, and they had to spray the floor with anti-static stuff to keep it from happening. Like everybody else, I have no control over it. It would be quite spectacular if we could, don't you think? [152]

Here we see an aspect of SLI we shall consider in greater detail in chapter seven, that many SLIders also affect other appliances besides streetlights. But they don't *all* do so – one of the many puzzling features we shall come across.

This SLIder is unusual in that her SLI functions in broad daylight. Streetlights being nocturnal creatures, few SLIders get the chance to find this out.

A Norwegian SLIder's experience brings out some other features of the phenomenon:

> One time I was driving in a tunnel in a traffic line (very slowly) and five lights went out one after another as I drove beneath them. [214]

This tells us that the lights do not need to be in the open: a tunnel is fine, so long as the light can "see" the SLIder. The incident is a classic instance of lights going out in sequence, and the fact that he was driving in a traffic line emphasizes that it was he, and none of the others, who was affecting the lights.

Most SLI events are spontaneous, but sometimes SLIders can

use their will power to dramatic effect:

> *A housewife from Birmingham, England:* On a visit to Ireland I was walking with some members of my family, and we were discussing paranormal happenings. I was telling my mother about the streetlights that I turn off outside our home, and she was laughing at me while I was talking, then a streetlight went out. She was a bit taken aback but said it must have been coincidence. Well, I decided to show off a bit, so I said "Watch this then." By this time we had walked on a few yards and the light had gone on again (as they usually do for me). As I walked back and approached the light it went off again. Well, she was incredulous. I was delighted it happened as it bore out my story. When we got back to my sister's house a bulb went as we walked in the door. You can imagine the reaction. [82]

Some people do SLI only during a brief period of their lives:

> *An electrical design engineer (English, female):* In November 1992, it was the first anniversary of my mother's death. After work I have to walk about a quarter of a mile from the British Rail station to the carpark. The most direct route is down a narrow, well-lit alley which leads to a flight of wooden stairs down to the beginning of the car park. As I descended the stairs and started to walk along the first section of the car park, one of the sodium streetlights illuminating the area winked out. This same light went out on the next four evenings, going out at a greater distance each time, the last two or three times seemingly

coincided with me thinking about it going out. I did not wait for the light to restrike. Each time I left it in an extinguished state. After five days, the SLI effect stopped as suddenly as it started, although nothing about the weather, time of day, temperature had changed dramatically that I can recall. [81]

An unusual detail of this account is that, on her later occasions, she thinks the extinction coincided with her thinking about it. As we have already seen, and will see many more times, SLI generally "refuses" to happen when the SLIder is thinking about it or willing it to happen.

This selection of reports brings out many of the typically recurring features of SLI, but we shall see that SLIders can pull many other tricks from their bag.

And the cause? A product engineer from the U.S. writes

> This has been happening to me for as long as I can remember. At first I thought it was a sign from God or some other spiritual being, but to my disappointment I found the SLI website and a possible scientific explanation. [216]

He should be so lucky! What are we to make of this activity, which our correspondent from Arroyo Grande rightly describes as "ridiculous," which seems to serve no purpose, which benefits no one, and which is a tiresome nuisance for civil authorities (though not a very serious one)? And let's add: which doesn't seem to add to our knowledge of how the universe works or how people behave? Which appears to happen out of the blue, with no rhyme or reason, no deep emotional repercussions or ill effects? With no meaning, so far as we can tell, yet which, because we have been brought up to believe that *everything* has a meaning, challenges us for just that reason?

If anything like this has ever happened to you, you are probably curious to know what is going on. If it has never happened to you, you may be curious to know why other people think it happens. If you are of a skeptical cast of mind, you will be intrigued by the whodunit aspect, and look for simple, natural explanations to dissolve the mystery. If you are a tease-the-experts kind of person, you will be entertained by something that seems to be happening even though conventional science says it can't. For SLI is a fascinating puzzle, as challenging as a Rubik Cube and at the same time as addictive as a detective story.

People experience SLI in many different ways. For some it is a frequent occurrence, for others it is rare, even once-in-a-lifetime. Sometimes only one light is affected, others affect a sequence. The SLIder may be walking, cycling, or in a car. She may be close to the light or at a fair distance. He may be alone or with others. The only common factor is that a streetlight is somehow affected, seemingly by the approach or the presence of the SLIder.

SLI is not a local phenomenon: it happens the world over. We have reports from the Americas and the Caribbean, Western and Eastern Europe, Scandinavia and Atlantic Islands, Asia and Australia. The similarity of these reports shows that SLI follows much the same pattern wherever it occurs. And whether in Boston or Budapest, Manchester or Melbourne, the same question arises: what's so special about streetlights?

Well, perhaps they aren't so special. SLI is only one of a range of experiences that take place in the borderland between the physical and the mental. What is happening is unquestionably a physical effect – the switching off or on of an electrical gadget – yet at the same time it seems to be a mental effect, in that though this switching happens spontaneously, it evidently requires the presence of an individual human, often in an exceptional state of mind. In chapters seven and eight we shall be considering other kinds of extraordinary behavior; but SLI is unique in that it involves highly visible objects, located in public places, where any kind of tinkering with the object is so improbable that we can reasonably ignore it. It isn't something that happens in the privacy of a home or the isolation of a laboratory. SLI is not hearsay or legend or rumor. It isn't something that happened to a friend-of-a-friend. SLI happens to real people with names and addresses – and is seen to happen.

In this, SLI is unique and can reasonably be regarded as a phenomenon in its own right. Confirming this, we have the testimony of hundreds of people, whom we have no reason to doubt. We shall be looking more closely at the people who do SLI in our fifth chapter, to see if we can identify what makes them special. Even if it should turn out that a few are telling lies, or making up stories, or exaggerating, the fact that so many people have put themselves through the trouble of reporting their experience – people from all walks of life, from many countries – makes it hard to believe that they are deceiving us or themselves. They could be wrong in the way they interpret their experience, and this is something we shall have to consider, but that *something* is happening, there is no doubt. Even if the witnesses are deluded, such a widespread delusion would be of interest as a social phenomenon.

That said, we have many reasons to think that those who report SLI are *not* deluded. What we are studying is indeed a scientific enigma.

But *which* science? What happens to the lights is surely a matter for the engineer or the physicist, but what causes it to happen seems rather a matter for the psychologist. Could it be that this is precisely the message of SLI: that it is trying to tell us that the world of mere matter and the world of the mind are not as distinct as we think? We do not have to suppose that SLI is a psychic happening, or that it is in any way supernatural. But we do have to consider the possibility that SLI is showing us a wider potential of the mind, a deeper interaction between mind and matter. In our final chapter, we shall try to assess the significance underlying what happens when someone walks down a street at night and the lights go out.

Chapter Two

Easy answers, hard questions

Science is the art of demystifying mysteries. Faced with the mystery of SLI, scientists – and indeed any thinking person – will, quite rightly, look for simple, logical explanations. The more so, because streetlights are man-made objects. Confronted with the mysteries of Nature, puzzlement is forgivable. But when it's something we ourselves have designed and constructed, there should be no mystery.

Yet mystery there is. So our first question must be the obvious one: Is there really a phenomenon, or could it be an artifact created out of people's imagination? Because SLI runs counter to known ideas about what is possible, and because it has only recently been noted, it is understandable that doubts exist, first, as to whether it happens at all, and second, whether what people *say* is happening is *really* happening.

A classical philosophical question is: Does such-and-such a phenomenon exist when no one is looking at it? This is a fair question for us to ask: Does it happen when there's no one around? But as with so many philosophical questions, there is no way we can answer it. Even positioning cameras to monitor the lights would fail because the human presence of the SLIder is apparently a necessary part of the puzzle. No SLIder, no SLI. Moreover, with so many people claiming to have experienced SLI, it is hard to deny that SLI happens. The reports we looked at in our scene-setting opening chapter show that people had an experience *of some sort*; the accounts themselves are clear, explicit, and indisputable. But what are they accounts *of*? Could it be a perfectly natural event that is being misinterpreted by the people who have it happen to them? Skeptics point to the witchcraft manias in distant centuries, and to the flying saucer mania of more recent times, as evidence that just because a great many people believe something to be real, doesn't make it so.

Such an attitude is right and proper. If there is a natural explanation, we should go along with it. Most objections raised by doubters can be easily answered, but they are useful in that they highlight the strangeness of the SLI phenomenon and draw our attention to possible – and impossible – explanations. So what are the possibilities?

Coincidence

> *A scientist from London, UK:* Having a mathematical background (MA, Cambridge) I concluded that it was massively unlikely that my SLI, which occurred fairly regularly while I was walking across Barnes Common [in southwest London] to have occurred by chance, especially as I was always the same distance (5 meters or so) from the light when it failed. I couldn't believe this was typical of the failure rate of lights at all. I thought about the possibility of me shaking a loose connection etc. and eventually concluded that I must be emitting some sort of aura – possibly electromagnetic radiation. Having studied physics I could see how a gas discharge tube light could have failed due to some sort of electromagnetic radiation interference, if the arc was only just powerful enough to light the light. [154]

Streetlights aren't immortal. Like the rest of us, sooner or later they die. So simple coincidence is the first explanation that comes to every mind; someone just happens to be passing when a streetlight breathes its last, and he imagines that one event is related to the other. The argument was neatly set out in a letter from a reader in Austin, Texas, to the British journal *Fortean Times:*

> I think there is a rather mundane explanation for SLI. Basically, the lights are cycling off and on, and a few of us have become unconsciously attuned to this behavior.
>
> High-intensity discharge lights, such as the high-pressure sodium ones widely used in streetlights, cycle off and on as they near the end of their lives. They can remain off for several minutes while cooling before turning back on. Most people do not notice the usually subtle effect of a single light turning off (or on) in their field of vision. However, some people, for whatever reason, notice it for the first time and may even see the element still glowing to confirm that the light was on just a moment ago. They then notice it on subsequent occasions and after a while some part of their unconscious becomes very adept at noticing it. They of course cannot notice a light transitioning when they are not around to see it and do not notice it when the light is far away because the effect occupies a much smaller visual arc. It therefore appears to happen only when they come near it. Moreover, the relatively long period compounds the illusion because they assume that the light has simply turned off. These people – myself included – are special due to their awareness of – and not their effect on – streetlights.[1]

Nicely argued, and perhaps this does account for a few cases. There are, after all, millions of streetlights illuminating the streets of the world, and billions of people passing them continually. Simply by the law of averages, it is argued, people are likely sooner or later to walk near a dying or malfunctioning light, whereupon they suppose that their physical presence is somehow responsible for its demise.

But the explanation becomes less plausible when we find that the same people have it happen to them time after time, sometimes throughout the course of their lives, and that a great many don't simply notice *one* light, but several, sometimes in sequence, sometimes at random. And that's before we take into account factors such as the state of mind of the individual, which we shall consider in chapter six. Add these factors together, and we realize that the odds against coincidence are astronomically high. On a personal note, I walk to my place of work in darkness throughout half the year, passing some twenty lights. I have only *once* had one of them go out as I approached, and it was clearly a failing light. After a couple of days it was replaced, and it never happened again. When we find a Canadian SLIder reporting "I have extinguished up to twelve streetlights in one evening: my average is two per week," [83] unless we suppose that Canadian streetlights are particularly frail, it is evident that something more than coincidence is taking place.

Consider these cases:

> *After recounting two matter-of-fact SLI events, this accountant from Harpenden, UK, reported a third:* On 1-9-1990, after we had discussed the matter, I was walking towards Baker Street [London] following a meeting. I was talking to Steve G. [a hard-headed mutual acquaintance]. I was discussing some strange events and Steve was rationalizing them to counter my suggestions. As we passed a streetlight on the corner of the Outer Circle [in Regents Park] and the main road, it suddenly went out. I said to Steve, "Witness that – I have to tell Hilary Evans about this" (and I briefly explained my previous SLI experiences). Steve countered by explaining that it was perfectly normal for streetlights to

"rest" for periods and that I should make nothing of it. I did not counter the point, which I am sure is probably correct; however, as we passed the next light in the street, that went out just as we walked past it. I asked Steve to witness that also. On neither of these occasions did I feel interactive with the lights, or even the cause of the incident. [8]

Perhaps related to that is the following:

> *A nursing student from Texas:* A friend and myself were talking about my ability to put out lights; he did not believe me, so I concentrated and nothing was happening. As soon as I gave up, 7 or 8 in a row went out. [10]

An English student who does SLI also reported this:

> Recently I had just checked out a book from the local university library. While I walked the short distance from the circulation desk to the magnetic exit arms, I suddenly knew I was going to be stopped at the gate, and I was. The device that demagnetizes books when they have been properly checked out had not worked, and the process had to be repeated again before I could get out. [13]

Simple coincidence? Maybe, but when it happens to someone who also affects streetlights, we start to wonder. And here are some more cases where coincidence seems an inadequate explanation:

> *A microbiologist from Norfolk, UK:* The first possible SLI occurred when I was in an extremely

emotional state, while I was driving to a police station. I was very worried about a loved one, and spontaneously asked, "Just give me a sign that she will be ok." At that instant a streetlight over the car burned out. Everything turned out ok. I am not sure who it was that I was asking for a "sign." [48]

A lady from Texas: As I was walking my dog one evening, after the streets were deserted and quiet, when an uneasy feeling of not being safe came over me. I was at that moment wondering if I should continue in the direction I was going. I stopped in the middle of the street and as I glanced down the street every single light for the neighborhood block went out followed by a small explosion and sparks from what I thought was a transformer at the other end of the street. It was the most odd thing to be immediately certain that this was NOT the way I wanted to go. We turned round and went straight home. [18]

A motion picture executive from New York: Two examples may relate to one another. The first is one of the few times I was able to *look* at a streetlight and it went out. I was walking to my car one evening from work and I looked at the streetlight ahead and I asked the light if I was going to move to Los Angeles. The light was already OUT and one second after the question the light went ON. I was bedazzled. When I got in my car, though, still looking at the light, it went OUT. I still do not understand this occurrence; maybe it was a sign. I do not know.

The second experience was also connected

to Los Angeles. I went to LA on vacation after being laid off from a movie company in NY, and the streetlights were going out left and right. My friend was in the car and noticed it as well. Jokingly he said I brought a lot of negative energy with me after being fired. [31]

A teacher from Paris, France: In 1980 or 1981 in Sofia, Bulgaria, I was walking to the Cité Universitaire, and stopped to drink at a little fountain. At the moment I stopped to drink, the streetlight overlooking the fountain went out. I laughed to myself, and the next day I told the story to a friend. We were in a bus, and as we passed by the same spot, I showed her the light which had gone out. At the very moment, it went out. It was one of the rare occasions when I have had a witness, the only occasion when I've been in a vehicle when SLI occurred.

On another occasion I was walking along a street undergoing repairs when a SLI occurred. A few days later, in the same spot, I remembered, and watched the light to see if the same thing would happen. Nothing happened. I turned my thoughts to something else – probably how not to fall into a hole in the road – and the neighboring light went out. My feeling at the time was that it is useless to think of SLI and hope to provoke it. It happens when it chooses to happen. [40]

A woman from Toledo, Ohio: The other day my brother was making fun of my supposed powers. He was laughing, saying that my experiences with the phenomenon were a bunch of hula-hulo. [sic] I ignored him, thinking to myself I know what I know, and his doubt could be attributed to being a younger inexperienced sibling-rival who possibly could be jealous. Minutes later, all the power – lights and everything – went out for 15-20 seconds, then came back on. I laughed heartily and exclaimed to him how wrong he was. He was flabbergasted and said to me, "Do it again." [42]

A photographer from New York, visiting the UK: Another amusing incident occurred in Exeter Cathedral when our choir was in residence for a week. As we were standing in the Close, one of the great floodlights went out. I had been telling one of our choirboys about my funny street phenomena, and assured him the light would come back on – which it did. He avoided me for a while! [44]

Each of these narratives is unsupported and rests entirely on the say-so of the SLIder. Scientifically, they are of minimal value as evidence. Yet if we are to arrive at a fair verdict on what is happening, we must take such accounts to be authentic experiences *of some kind.*

This is an appropriate moment to relate another personal incident. I was working on the original notes and reached an account, narrated in chapter seven, where a SLIder claims he nearly always knows when his computer is going to "die" on him, when my own computer – which had given me no trouble for a year – suddenly died on me. At the time I wrote: "Asked to choose between cosmic joke or simple chance, I have to admit I go for the joke, though whether the joker is some mysterious demon or my own equally mischievous subconscious, I am not prepared to say."

But that isn't the end of it. Some months later I had reached precisely this point – considering the likelihood of coincidence

– in a later draft, when the coded house alarm, *which I had not set*, went off spontaneously, along with two other incidents involving misplaced objects for which we could find absolutely no explanation. We shall be examining this kind of synchronicity in our ninth chapter.

Expectation

One or two SLIders have claimed that they sort of "know" when they are about to kill a light. If this were generally the case, it might be a valuable clue, because it would tell us that somehow the individual *expects* the interference to occur, which is tantamount to saying that it is pre-ordained in some way. That could tell us that the SLIder *intends* to blow a light, which every SLIder denies and for which there is no evidence whatever. Or it could simply mean that the light is on the verge of blowing anyway, for whatever reason, and that the witness is somehow aware of this, in the same semi-mysterious way that an experienced mechanic, when you drive into his garage, can tell you to get out your Mastercard because he knows from the sound of your engine that something is about to go wrong with it.

But that's too many "somehows" to be convincing, and though it's an interesting line of thought, and worth bearing in mind, it is no sort of an explanation as to why the light extinguishes and why the individual thinks – *after it's happened* – that she is responsible. In any case, it is liable to all the same weaknesses as any explanation based on chance coincidence.

Faulty lights

If a light is not in its death agonies, it may nonetheless be suffering from some lesser non-terminal ailment. Closely linked to the coincidence hypothesis is the suggestion that the lights affected are faulty anyway. This is very likely true. The fact that SLIders turn off some lights and not others suggests that the ones they turn off are more fragile, more vulnerable, weaker in some way, by comparison with others. Quite a number of reports support the view that this is the case. For example, consider this instance from Leicester, UK:

> This seemed to be a very regular occurrence. I think I first noticed this because I lived in a tree-lined avenue which links the Leicester University to the town centre. In between these trees were Victorian lights. (I am not sure whether they were originals but I could quite believe they were, certainly the avenue was extremely old even with its name "New Walk.") In the evening when I was going to and from the town centre at least three of the lights would switch off as I approached and then relight once I had walked past. It became quite noticeable because although the lights lined the whole avenue, it was always the same three that were affected. I did not pay too much significance to this and certainly would not have put it down to any paranormal activities. I simply explained it away as very similar to incidents which used to occur whilst I was wearing a watch which would stop working if I wore it for a while – if I took the watch off it would function as normal. [110]

A fair deduction from this is that the three lights affected – always the same ones – were weak or otherwise vulnerable compared with the others. That tells us that weak lights are more easily affected than their more robust companions, which we might have figured out for ourselves. But it does not explain why this individual regularly caused them to extinguish and then relight, when other passers-by didn't affect them.

No SLIder has ever reported turning off *all* the lights she passes, and this is surely a very significant fact. It tells us that *some* lights respond to SLI while others don't, and this can only mean that they are either defective – though not to the point of going out of their own accord – or fragile in some respect, which could be a manufacturing weakness or simply the consequence of normal wear and tear over a period of time. Since the Leicester lights came on again, however, the defect was evidently not serious enough to stop them working altogether. And if they were continually being extinguished by every passer-by, they would surely have been replaced by now. So any explanation based on a congenital weakness of the lights is ruled out; the career of a streetlight is not a Darwinian struggle for survival of the fittest.

Moreover, we have cases – for example, number 196, which we consider in the next chapter – where an entire row of lights are extinguished in sequence. This may mean that all are equally robust/fragile because they were all installed at the same time, or that the SLIder is able to extinguish the fragile and robust alike.

In January 1990 Robert McMorris of the *Omaha World Herald* devoted three successive issues of his regular column to SLI.[2] Looking for a natural explanation, he cited Allen Klostermeyer, a representative of Lighting Specialists Inc, who pointed out that:

> When "a sodium (amber) bulb nears the end of its useful life, it may go off and on in sequence." When one of them starts to "die" it requires more voltage. This will cause the light to go off temporarily; when it cools down, it will come on again for a while. Eventually it will die completely.

The proposition is that the supposed SLIder just happens to be passing the light when it is behaving in this way, and surmises that he is responsible. Possibly this is what happened to this woman researcher from Boulder, Colorado:

> I drove to a nearby town to attend a meeting. As I was early I parked on the street to wait. There were streetlights lining both sides of the street. I had

> been sitting in the car for about 10 minutes when the streetlight on the opposite side of the street (and more than 50 meters away) suddenly went off. I watched in total fascination as it came on again, then popped off again. After about a dozen times I decided to time it. It wasn't random, but it wasn't cyclical either. The time between going off and coming on seemed to vary from 45 to 55 seconds. There seemed to be no relationship to passing cars, as it very often popped off when there was no car near it. Anyway I left to attend my meeting, and when I got out, that streetlight was still bonkers. Perhaps it was defective, but I suspect it just became defective because I was in the neighborhood. [13]

The distance from the observer to the light strengthens the view that it was just coincidence; on the other hand, the fact that it started 10 minutes *after* her arrival could be significant. And what made the observer think she was responsible? The answer to that is that this woman has a long history of SLI experiences of various kinds. Even if we disregard this particular instance as coincidental, the fact that it happened to someone with other SLI experiences requires us not to dismiss it outright.

Richard Wiseman, head of the Department of Psychology at the University of Hertfordshire, England, is one of the few scientists to give SLI his serious attention. He was kind enough to send me the result of his calculations based on a mathematical concept, the Law of Very Large Numbers. This law explains many seemingly extraordinary happenings by demonstrating that, taken a large enough scale, they are not so extraordinary after all.

"Assuming," Wiseman wrote, "that people on average take 3 seconds to walk under a street lamp, and walk under approximately 15 per night each night of their lives, and that they do this for 50 years, each person will spend 3 x 15 x 365 x 50 = 821,250 seconds walking under street lamps in their lives." He next proceeded to set the number of people in the world against the number of lights, and the fact that sooner or later a lamp is going to reach its expiration date. In the end, he reckoned the chances of a person passing a dying lamp – or even two or three – sometime in the course of his or her life, were much greater than you might expect.

That may well be so; I am not disputing his figures. But then we come up against another mathematical constraint: *averages*. The Law of Averages is a wonderful thing; it means I can walk into a clothing store and come out with a pair of pants that fit me, near enough. But let a 7-foot Basuto or a 3-foot Pygmy walk into the store, and they will need to be shown to the custom department. And so it is with Wiseman's figures. If every SLIder put out the same number of lights – say one a week, Wiseman's figures would be dandy. But we've already seen enough reports to know that there's just no pattern: some do one or two. Some do a whole batch at a time, then zilch. Yet others do it all their lives, slaying them by the score. Good try, Professor, but in the face of this diversity, the Law of Averages makes your Law of Very Large Numbers irrelevant.

Types of light

Some would-be explainers suggest that only a particular type of light is affected, but we shall see that this is simply not the case. Most SLIders don't know much about streetlights and aren't able to distinguish one kind from another, but some are better informed and specify what kind they affect. Thus an English electrical design engineer notes that the lights she extinguishes are the sodium kind. [81] Moreover, some SLIders specifically state that they influence more than one kind of light. We don't have any reports relating to the old incandescent lights, but there aren't many of those still around for them to zap.

And this is before we take into account the many reports from SLIders who affect other types of appliance, which we shall consider in chapter seven, "Radios to railway crossings."

Car and bicycle headlights

Several critics have suggested that vehicle headlights could produce a spurious SLI-effect. American investigator Loyd Auerbach has dogmatically proposed:

> Almost certainly what was happening was that headlights of passing cars were being reflected into the photoelectric cells of the lights (the electric eye that automatically turns streetlights on and off as the sun sets and rises).[4]

This is of course nonsense, for if it were so, streetlights would be continually switched off by passing cars. Well, no; we could perhaps surmise that the headlights may have been pointing in just the particular direction required to strike a photoelectric cell mounted on the light – in other words, it's misbehaving headlights rather than misbehaving streetlights. Whether this could possibly happen is open to question, and if it were the case, the offending cars and bicycles would be causing havoc wherever they went, unless of course we take it a step farther, and suppose that what is involved is a fortuitous combination of misaligned headlights and defective streetlights. Which is frankly ridiculous.

One who at first was inclined to take this suggestion seriously was an informant from Exeter, UK:

> My wife and family always joined me in joking about this apparent paranormal ability of being able to switch off or on light standards by merely passing them. Although seriously I at first attributed it to coincidence, and assumed everybody else was getting caught out and imagining themselves latter-day Uri Gellers. But I soon rationalised that were this the case, then the streets of Exeter and its suburbs would be flashing on and off like Christmas trees. I then

> turned my attention to the fact that the effect was more frequent while I was driving, both in my motorcycling days and more recently in my car, and thought that perhaps it could be a reaction to misaligned headlights confusing the photoelectric cells that control the lights, particularly as it is not uncommon for rows of three or four lights to switch off as I pass. Logically, this still provides the "safest" answer to this puzzle as long as one is willing to believe that three separate motorcycles, six different regular cars, and many occasionally driven vehicles, all have headlights that point too far upwards, but only between MOT [Ministry of Transport] tests. I hope you will agree that this perhaps stretches logic beyond even a Vulcan perception of the discipline. [175]

But we do not have to resort to such possibilities because the headlight theory is quickly disposed of when an informant from New York tells us explicitly, "It happened when I was walking, driving, or riding my bicycle." [28]

Chapter Three

How to switch off a light

A young man from Dublin, Eire: I was leaving a friend's house to go home to my own, about 11pm at night in early summer. My own house is about a mile from his with a straight road that jinks slightly to the right and then continues as it crosses a main intersection. As I start walking up the road the streetlight about seven meters ahead of me goes off. I don't pay any heed and continue on. As I get about the same distance from the next one, it too goes out. Okay, I pause and think to myself, "Two in a row? Bit of a coincidence," and carry on. Yes, you guessed it, same with the next one. This time I stop and look back at the other lights: they've come back on. I start walking again and this happens all the way home except at an intersection. I get seven meters from the light, it goes out. I get about three to four meters past it, it comes on again. Needless to say I am walking quite fast and a little freaked out by the time I get home. I tell my friend about this the next day; we are both at a loss to explain it and eventually say maybe it was just weird coincidence (very weird!). As time passes I begin to think I may have imagined it or maybe had a really vivid dream. And no, no substances were taken, legal or otherwise.

About six months pass and the incident is pretty much forgotten. Unfortunately at this time my father suffers a major heart attack and is taken to Blanchardstown Hospital, Dublin. All

> the wards are in different buildings to reduce the chance of infection. My father wants me to get him a paper from the hospital shop in another part of the grounds, and the same thing happens again all the way from the ward to the shop and back. Seven meters from the light, it goes out. Four meters past, it comes on again. Totally freaked out again but I can't explain to my father why I am white as a sheet when I came back. He doesn't need me telling him scary stories. Hasn't happened since. [196]

What happened to that young Dubliner was impossible by any standards. Impossible by what he did, impossible by the way he did it. Yet it happened. Somehow, between him and the lights he passed along the street or in the hospital grounds, a process took place that contradicts our conceptions of what can happen and what can't.

Setting aside for the moment those conceptions of what can *actually* happen, how can this young man *conceivably* have this happen to him?

The one thing we can be sure about is the light itself. We all switch lights on and off many times a day. They are man-made objects, we know what they are, we know what they are designed to do, and at least some of us know more or less how they do it. It is easy to switch off the kitchen light because that is what it is designed to do when you no longer require its services. Streetlights, on the other hand, are emphatically designed *not* to be turned off by the likes of you and me or our friend from Dublin. The designers made it next door to impossible to do. We could say *impossible* – except that this Dubliner and a few others have found a way to do it.

So how do they do it? The Dubliner's experience is a classic example of SLI. It presents the enigma in simple, unambiguous, explicit terms. So we can ask simple questions like: Did the lights *choose* to switch off just as he approached? Surely not; streetlights don't *choose* how to act, they don't have any choice in the matter, they do what their designers designed them to do. Well then, did the young man *choose* to turn the lights out, the way you and I decide to switch off the kitchen light? Surely not, in any conscious way. Yet somehow an action – a mutual transaction, almost we might say an *agreement* – took place between unchoosing him and the unchoosing lights. And the lights went out.

But who or what agency negotiated that transaction? Before we've finished, we will have to consider all sorts of possible candidates, ranging from the young man's subconscious mind to angels, poltergeists, and Cosmic Tricksters. But whoever or whatever brought it about, we start from the incontrovertible fact that a streetlight is an inanimate piece of machinery, and to make it do anything, the angel, poltergeist, or whatever has got to monkey with the mechanics of the appliance.

So, how can a light be switched off, outside the protocol of its regular, normal, controlled dawn-after-dawn extinction?

We shall see throughout our inquiry that there are almost as many variations on the Dubliner's experience as there are SLIders with tales to tell. If every SLIder told the same tale, life would be a lot simpler. But they don't, which makes things more interesting, but also makes them more puzzling. Most SLIders turn the lights off – but some switch them on; many do both. Some lights go out as the SLIder approaches, some only when she is directly underneath. Mostly the SLIder needs to be pretty close to the light, but sometimes lights go out on the far side of the street or even further away. Some lights come on again as soon as the SLIder moves away, others only after an interval, yet others not till the following evening. Usually it's simply a matter of going on or off, but sometimes they flicker, sometimes sparks fly.

The SLIder can be walking, cycling, or driving a car. The

most frequently reported are driving or walking to work, but a Devonshire informant, when she was living in the small town of Katrineholm, Sweden, used to cycle to work: "At first I thought it was coincidence but it happened so regularly that I became aware of it." [173] She continued to have this effect after she came back to England. This brings out another feature of SLI: it travels. An informant from New England wrote:

> While living in LA, I started noticing that streetlights would go off or on when I drove underneath them. I moved to Chicago, didn't have a car, walked a lot. Streetlights would go off/on when I walked underneath them. In 2009 I moved to Connecticut. Same thing. [215]

One thing we can be sure of is that SLI happens only when someone is close by. Whether or not angels or poltergeists are involved, SLI is a dialogue between a light and an individual. And mostly it's an involuntary dialogue. Though we shall see that some SLIders have successfully willed a light deliberately to go out, generally it's spontaneous, something that just happens.

All of which is rather unhelpful – except that in a negative kind of way, this diversity tells us something important about SLI. It tells us that there's no use looking for a single, simple solution. It's not a case of if you do such-and-such, then so-and-so will follow. SLI is flexible and adaptable, unpredictable and as many-faced as a comic actor. Which perhaps is what it is…

Why streetlights?

As we shall see in chapter seven, streetlights are just one of many appliances, devices, and instruments that may be affected by SLIders. Despite the competing attractions of cars, computers, and cash registers, streetlights continue to be the favorite targets. If we knew why they are singled out, it might be easier to understand what is happening. But as investigators, we should be grateful, for they are admirable subjects for study. They are out there in the open, in the public domain, for all to see, as opposed to being tucked away in a private, domestic or laboratory, setting. This means they are not likely to be tampered with; forget naughty boys, vandals, or deliberate mischief-makers. The chances of being seen are too great, and besides, few people except lighting engineers would know how to go about it.

Another thing about SLI: what happens is "clean" – there are no side effects, no after effects, no angst or trauma, no bills to pay, no mess to clear up. When you think about it, this combination of virtues is well nigh unique. Indeed, it isn't altogether crazy to consider the possibility that streetlights are selected precisely *because* they make such ideal targets.

A would-be SLIder's introduction to streetlight technology

If you are going to be any good at switching off streetlights – and some SLIders are *very* good – you need to know how they work. Not many people do. *I* didn't, so I had to recruit a number of experts to help me write this section of the book. And they didn't always agree with one another.

The first thing you will need to ascertain is what kind of light you plan to interfere with, for there are several kinds in use. Most likely it will be a High Density Discharge (HDD) light, of which those most commonly used are High Pressure Mercury (the greenish-white ones), High Pressure Sodium, or Low Pressure Sodium (the amber ones that make your loved one look as though he/she's at death's door). They all operate in much the same way. Fluorescent and incandescent lights are also occasionally seen, but their low output means that you will not often come across them outdoors, though of course they are widely used for interior lighting. Many SLIders specify that they affect HDD lights; other types are rarely mentioned.

HDD lights operate by passing an electric current through a liquid metal (e.g. mercury, sodium, or tungsten), which first has to be pre-heated by a starter circuit to form a gas vapor. Once the vapor is heated up to start-up temperature, the general light circuit takes over, and so long as adequate voltage is maintained, the light will be alight. There is a period while the lights first come on, during the heating process, when they glow but are not fully lit. Once that point is reached, there has to be a continual flow of electricity into the light for it to stay alight; interrupt that flow, and the light goes out.

The phrase "adequate voltage" suggests a possible chink in the light's armor. Too low a voltage, and it will not be sufficient to keep the light alight. If the voltage should drop below a specified level, the light's internal resistance will overcome the circuit and the light will cease to function. Too high, and it will cause a surge that will automatically switch off the light to prevent damage. Needless to say, the light is designed first to reach an adequate level and then to stay there, and built-in protective devices ensure the voltage is maintained at the required level. The people who install the systems don't want you messing about with them, so access is difficult to anyone except an electrical engineer. Few SLIders, if any, are electrical engineers possessing the necessary know-how.

If an HDD light is turned off during operation for any reason, it can't come on again right away. There has to be a re-strike period, while the light cools down before the discharge process can start over. Normally it switches on again automatically, but nothing will happen for a couple of minutes or so until the light reaches the proper level. So there is no question of the light coming on again immediately to full illumination; it can't.

The electricity is supplied to the light from a feeder pillar, which it shares with other lights on the same circuit, in parallel, but to which each light has its own independent connection. Theoretically this is a likely target for your SLI, but the supply circuit being demand-oriented (i.e. responding to the power requirement at any time), it possesses built-in protection against undue surge or reduction. Moreover, it is concealed and insulated.

If you were to interrupt the *main* power supply, this would result in *all* the lights in the circuit being extinguished. This is rarely reported by SLIders, and those few reports are questionable. Generally the lights are extinguished *individually*. Even if several are extinguished, as in the Dubliner's case, each individual light must be affected *separately*, in succession. So if the power supply is your target, you must be prepared to attack each individual power supply connection as a separate action. To do this while you walk or drive along a street is a formidable task.

But if it isn't the power supply, what alternative targets do you have? For example, every light has a fuse that is blown if a component becomes defective or some malfunction interferes with operation. The fact that the lights usually come on again after a brief time – generally within minutes – implies a self-rectifying process that operates automatically once the fault has been identified and set right. It is hard to see how you could do this. Moreover, fuses, like the power supply, are concealed and shielded, and as with the power supply each one would need to be affected individually as you approached the lights in sequence. Another very demanding task.

The alternating current powering the light is of such high electrical potential that it is unlikely to be over-ruled by another force of substantially less potential. To interfere with it, you would need to exert a very strong force. True, we have no idea what limits, if any, to set to the SLI-force. But it is hard to believe that your body could send out a truly massive force without feeling drained or exhausted, whereas in practice very few SLIders feel any sensation of depletion at all, either at the time or subsequently.

How about sensors? Lights that are controlled by light-sensitive sensors, which tell the light it's getting dark and time to go to work, would seem a likely target. The SLIder has only to make the light think it's daylight, and off it goes until he moves away, when it will realize that it is still dark, and comes back on again after the usual interval. Lights controlled by independent light sensors are not all that common; often the lights are operated as a group, but reports of several lights extinguishing together are very rare. The sensor is positioned on top of the light, out of sight of the SLIder; we do not know if this would be a serious obstacle. If it is essential for the SLIder to be within sight of the sensor, this is hardly a viable procedure – we have no reports of SLIders clambering to the top of light poles and few have the necessary agility. On the other hand, if all that is required is that he should be in the neighborhood, then indeed some method of over-riding the sensor's built-in reluctance to operate in the absence of light just might do the trick. But you'd have to work out how.

The same objection applies to a relatively new technical development, the motion sensor. What happens here is that a sensor, mounted on the light, detects the approach of a pedestrian, cyclist, or motorist, and switches the light on for a sufficiently long period to light the way. This would appear to be ready-made for the SLIder. However, though the technology exists – like many householders, I have them on the front and side of my home – it is not yet proven how viable it would be to employ them on a wider scale. The city of Toulouse, France, is the only sizeable urban area to try it out; it remains to be seen whether the saving in power justifies the cost of installing and maintaining the equipment. And there are other questions: will every passing dog, fox, rabbit, or hedgehog trigger the light? Will people tolerate the continual off-and-on switching? What about naughty boys and pranksters? So for all practical purposes, for the time being this way of putting out a light is unlikely to be found on a wide scale. Should this come about, of course, it will be difficult to distinguish SLIders from the rest of the population.

Another possible target is the switchgear that turns the light on and off. There are three different systems. In the first, a circuit of lights is switched on/off from a main control center. In the second, each light has a timer, pre-set for the light to come on/off at a given time, again regulated by a central control. Or each light or group of lights may be activated by a photoelectric cell that responds to actual conditions, bringing the light on when it is sufficiently dark, switching it off again when it gets light.

It is sometimes suggested that it is the photoelectric cell itself that is activated by the SLIder. Theoretically, assuming that each light has its own cell (not always the case), you could monkey with the light's operation by interfering with the cell, making the light think it is dark when it is daylight or vice versa. In practice, however, these cells are generally placed on the top of the unit where they will be most receptive to prevailing lighting conditions but where they would not be in your line of sight.

A further consideration is that it offends commonsense that the process of operating a switch should work both ways – that is, that if it (whatever "it" may be!) finds the switch is set one way, it switches if off, and if the other way, switches it on. Admittedly, commonsense is not something there seems to be too much of when it comes to SLI, but we should adopt a more sensible alternative if there is one, and in this case I think there is. Messing with the power supply makes better sense from every point of view.

This raises the question of how essential it is for the light to be able to "see" you. We do not have a single instance of a SLIder affecting a light when he is out of sight of it; consequently we may reasonably presume that this is an essential requirement.

It does not necessarily follow that the actual component that you affect must be visible to your eyes. But it must be within the field of your SLI-force. Your presence is required. But this is a very puzzling matter, since we have cases in which the SLIder is 50 meters or more from the light he affects. That being so, it would seem that your *immediate* presence is not necessarily essential. Yet many SLIders do not affect the light until they are within very close vicinity. Why this difference? We shall come back to this question of distance in the next chapter.

A further unresolved question is whether we are considering a force field that operates in all directions, or whether it is specifically focused on one particular light at any one time. Given that streetlights are generally placed a considerable distance apart, it is not easy to see how this could be tested. But the fact that SLIders hardly ever extinguish more than one light at a time suggests either some form of target-seeking or other selection process.

Could you interrupt the light's operation by affecting the internal gas vapor pressure or quality, or the internal fluorescent materials? One expert remarks that "of all the light's apparatus the gas is the weakest physical link," but that does not mean that it is a suitable target. Interfering with the gas would require an appreciable length of time, whereas SLI appears to be virtually instantaneous.

All in all, an interruption to the power supply seems your best option. In principle this can be done in either of two ways: by *reducing* the amount of electricity, or by *raising* it in a "surge." It will stay off until voltage is restored and the re-strike process is completed. Most lights come on again very soon after the SLIder has passed it, which suggests that the SLIder somehow interrupts the voltage flow temporarily. When the light does not come on again until the following evening, though, it is evident that something has interrupted the current more lastingly. This could be simply a matter of what it is programmed to do.

But just as there is a lower limit to the voltage at which the light will function, so there is an upper limit. If this critical level is reached, the circuit will kick out so as to preserve the light from damage due to operating at too high a voltage. Under normal conditions, an overload of this kind rarely occurs, because the circuit is designed to operate within defined limits, despite external conditions such as humidity or the ambient temperature.

So what this boils down to is that your most favorable option is to bring about a surge. How would you do this? You would have to somehow generate an electro-dynamic force within your biological system, and externalize it into the neighboring environment, where it will act on any appliance that happens to be both close to you and sufficiently vulnerable.

It won't necessarily be a light; the same modus operandi could interfere with the working of a computer or any number of other appliances such as hair dryers or cameras. However there is reason to think that lights may be particularly sensitive compared with, say, a washing machine. And there is the curious fact that some SLIders have no trouble with other appliances.

None of this will be done consciously. You couldn't do it even if you wanted to. The generation and deployment of the SLI-force needs to be an involuntary process carried out by your unconscious mind. To do SLI, your unconscious needs to be very clever indeed. For now that you have identified your target and chosen your method, all that remains for you to do is figure out how to generate the force field within yourself, and how to direct it at your victim. Then, hey, you're doing SLI.

Doing SLI

The preceding section has been a somewhat tongue-in-cheek exploration as to how SLI might be performed by someone sufficiently devious to want to do so. We have gotten some way towards understanding the fundamental mechanics that SLI entails. But it doesn't take us very far, because of course it is completely impracticable. A SLIder doesn't carry about with him the necessary tools and equipment, and even if he had them, he wouldn't know how to use them. However, even a first step is something to be grateful for. It gives us some insight into what might be happening in a simple straightforward case. Though as we are discovering, simple and straightforward are two things SLI rarely is. Before we are done, we shall see that things can get a lot more complex and contradictory. Also, to be sure, more challenging.

For starters, consider this: if the SLI-force extinguishes a light by overloading it with a surge of power, what is it doing when it turns a light *on* rather than off? The implication is that, whatever the light is doing, these SLIders make it do the opposite. What kind of force could do such a thing? It is as if a doctor cured people when they are sick, but made them sick if they are healthy. Take this case from a resident of Melbourne, Australia, who works in the film/TV industry, and tells us "this occurrence has been happening to me on and off now for the past 5-7 years":

> Earlier this week I was parking my car near a university in Melbourne when I noticed a streetlight go out as I passed under it. The following evening I was parked in literally the same spot, as I returned to my vehicle with a friend. The light in question from the previous night re-lit as I came within close proximity. [121]

The implication is that this man's SLI-force operates in two modes, one in which it turns *off* the light, another when it turns it *on* again. Given that he shut it down with an overload the first night, can we suppose that the light sat throughout the second day, waiting for him to come back and restore it to its normal state by giving it a second shot of the SLI-force?

Well, yes, perhaps we can. Maybe that's just what happened; maybe another burst of the surge of the same SLI-power that put the light out of action the first night was just what it needed to get it going again. Far-fetched? Surely. But it is at any rate a plausible way to account for the on-and-off paradox, which we shall look into more thoroughly in the following chapter.

Meanwhile the fact remains that both the Dubliner and the Australian did something that is inherently *impossible*. And in the Irish case, the impossible occurred over and over again, in a matter of minutes.

The folk who design streetlights design them explicitly *not* to do the things these SLIders made them do. They want their lights to keep on making our roads and pathways safe for us to get about. So what secret weapon do SLIders possess that enables them to out-smart the professionals?

All we have got so far is the working hypothesis – and let's be honest, that's all it is – that the SLIder uses some kind of force that overcomes the technological know-how of the streetlight designers. We're calling it the SLI-force for the sake of convenience, but that's just a label; we don't really know a darn thing about it. But it is as though the Dubliner and the Australian carried about with them one of those handy gadgets beloved of science fiction creators, enabling them to zap those horrid Klingons with a bolt of disabling force. From what we have seen of the streetlights' mechanism, it's hard to see how they could do this, but given that they *can*, they *do*, and then they move on and after a short interval, the light's self-relighting mechanism brings it back on again. Not always, but mostly.

As an explanation, this is pathetically inadequate. And it gets worse. For as we are about to learn, the diversity – we might say the *perversity* – of SLI is such that we need an explanation that will meet a range of behavior much more puzzling than that which the Irishman and the Australian met.

For one thing, apparently *all* kinds of light may be involved in SLI, irrespective of their technology. Metal vapor lights are most often mentioned, but then they are the most commonly encountered. Several SLIders have reported old incandescent and fluorescent lights, not to mention domestic lights of various kinds. This is a crucial point, for it confirms that it isn't some technical feature of a particular type of light that is responsible.

It is probable that the lights affected are those that are particularly fragile or vulnerable, but if that were the only factor, they would respond to anyone and everyone, not only to certain individuals. When considering the possibility of coincidence in the previous chapter, we ruled out malfunctioning lights on statistical grounds. A few SLIders have mentioned types of light that they do *not* affect. For instance, four have told us "never with tungsten," and one specified "never with incandescent bulbs or neon." [15] Which, like so many little observations we are offered, makes our problem *more* difficult, not less.

Although we have discounted the easy explanation that car headlights cause that SLI, there does appear to be some association with cars, as is suggested by this case:

> *Message from a Californian on a bulletin board:* I want to describe an occurrence that has gone on for about 9-12 months, and would like to see if anyone else has such an experience. For some reason, when I have driven under a specific light at night, it will go out! Now, this does not happen when I am 50 meters ahead or behind it, but when I am exactly underneath it,. I have a sunroof in my car so when it goes out I can usually look up through the sunroof and see it going out. This has occurred about a dozen times in the last year. I never try to make it go out. About 50% of the time I am not thinking of anything specifically, and am somewhat upset the other 50% of the time. [3]

This drew the following comment from another contributor:

> Since streetlights contain light sensors, when you go under them with a sunroof, or a certain angle of the windshield of your car, the reflection of the streetlight on that surface causes it to shut off because the light sensor thinks it is daylight.

Which is ingenious, but of course fails to explain why he doesn't turn off *every* light he passes under. Still, there are useful clues here. The fact that the light does not extinguish until his car – with its open sunroof – is directly below the light is surely significant. If the man can see the light, the light can see the man; so we may reasonably guess that it is not the car but the driver that affects the light. That said, the fact that the light goes out instantly when he is below it suggests an almost instantaneous effect.

Can we then infer that the light needs to have an *uninterrupted* view of the individual? In this instance, yes, it seems so, but generally speaking, surely not, unless every SLIder who affects a light while driving has the window or sunroof open. If so, it is rarely if ever mentioned, and we have the impression that generally the car is closed. After all, it's after dark and likely cold; Swedes and Norwegian drivers surely like the car good and warm. And it rains sometimes.

We shall consider other odd reports involving cars in chapter seven.

Could there be a complex physical explanation? A Hungarian physicist, following his own experience of SLI, wondered:

> In my opinion during such incidents some special, presently not known type of magnetic field is created around the body which has an effect upon the structure of the materials. Consequently their fundamental properties are changed temporarily, like their tensile stress, electrical conductivity, magnetic momentum, optical properties, etc. The same effects are detected in the case of "metal bending," or similar features are observed sometimes around ball lightnings. [4]

The informant (who incidentally is a world expert on ball lightning) recognizes that SLI, whatever else it may or may not be, is a *physical* phenomenon, and therefore the explanation must be, in part at least, a nuts-and-bolts one, irrespective of any psychological element. As an engineer, he proposes the sort of effect that would trigger such a process. This can only be part of the explanation, of course, but it reminds us that whatever SLI is, it is ultimately subject to physical laws. Even if those laws have to be revised in the light of SLI.

It's something for us to have got that far. But let's not be too pleased with ourselves. For what kind of force field is it that is active whether out in the open or enclosed in a car; which seems to be unaffected by distance so long as line-of-sight is maintained; which picks on streetlights, but which varies so strikingly in the other appliances it affects – or doesn't affect – and which fluctuates so bizarrely with how the SLIder is feeling this evening?

And how is the SLI-field created? Every human body possesses a force field; it comes with being a living person. But evidently SLIders have one that's different from most folk; if it was just simply stronger, we would simply have weaker and stronger force fields. This may indeed be the case, but it isn't borne out by the reports that seem to suggest that you either do SLI or you don't. Many reports include seeing other people walking under the same lights that they themselves have affected, yet not affecting them in the slightest degree. Not a flicker.

Can we even be sure that the SLIder generates his force within himself? We are assuming that such is the case, but we also see good reason to question the assumption. Could it be that the SLIder is unconsciously tapping into some cosmic power source in the air around him? Ask your friendly neighborhood anatomist what part of your biological make-up is capable of doing these things, and you'll get the same look as if you asked him to locate your soul or your conscience. Just the same, it bears thinking about.

Chapter Four

The many ways of SLI

SLI is in every sense an on-and-off affair. It is most noticeable when the light goes out altogether, and this is what most SLIders report. However, as we noted in the previous chapter, some SLIders turn lights off if they are on, on if they are off. We have put forward a possible escape route from this apparent paradox, but this doesn't satisfactorily explain the experience of a woman in Austin, Texas, who says "I turn streetlights on more often than off, I think. My friend Gary turns them off more often than on." [157] It would be nice to explain this as the Power of Positive Thinking, but it fails to account for the experience of a priest from Birmingham, England:

> The usual thing is for the light I pass under to start flickering madly and then resume its brightness only when I am further along the street.... I have noticed the recovery as the strangest part of the phenomenon. [123]

> *... and a woman from Forest Hill, London, UK:* At the time my husband had left me, I used to walk to the shopping area in Forest Hill quite regularly, taking the route by the station underpass. There were of course many streetlights on the way, but I caused a reaction with only two; they were the lights on either side of the underpass. If the lights were still out as I was on my return journey, they would switch back on. [136]

A SLIder from Dublin, Ireland, affected streetlights on the way to visit his girlfriend – "sometimes one or two, sometimes a whole row of them." He emphasizes that "they would only remain off whilst I was underneath before coming back on again." [153] So we are faced with the fact that in some instances at least, the lights misbehave only in the immediate presence of the individual, and the moment that presence becomes an absence, the light returns to the way it was before. Can we cling onto this as a general rule? Sorry, no. An informant from Los Angeles tells us:

> Sometimes they will re-ignite shortly after I have passed under them and at other times I will look back and they still haven't relit. [34]

A singer from the Isle of Skye, Scotland, decided to test her SLI ability:

> More and more streetlights began to go off as soon as I approached and passed. Then one day I tried walking past a particularly responsive streetlight just as an experiment. It went off. I walked back, it went on. I freaked out! Couldn't explain it and decided not to think about it too much because it was all too weird. [168]

An American now living in England first noticed the SLI effect when he was living in the San Francisco area, and it led him to possible explanations:

> I particularly noticed it when crossing the San Francisco-Oakland Bay Bridge. I have seen more than one go out during a single crossing of the bridge. The frequency with which I witnessed the lights going out caused me to reason thus: If everyone saw as many lights going out as I do, then there would soon be none on at all. Therefore there must be something about me that causes them to go out. ...

> It was after coming to England that I made another observation: the lights that I observed going out turn themselves back on! I think that all of the lights that I have seen go out have been mercury or sodium vapor lights which, after a cooling-off period, re-ignite. Therefore each one can go out several times each night without the risk of plunging the country into total darkness. [111]

Does the strength of the SLI-force determine how long the light stays out? Near-instant re-lighting happens, but so do intervals of every possible duration. A SLIder from Reno, Nevada, tells us: "I observed the lights operating normally about two hours later," [25] whereas several have said they came on again the following day: Hickory, North Carolina: "The lights were always on when I passed them the next day" [1] and Budapest, Hungary, "next day they operated perfectly again." [4]

Long-lasting or semi-permanent effects are rare, but a few have been reported. A student from California tells us:

> Another SLI event occurred after I told a fellow-student about the strange occurrences, including the streetlights, while walking across a large university campus. When we finally reached my truck, I opened the door, and the lights on both sides of the truck burned out. They made some crackling noise. The student was quite startled. We looked around and all of the hundreds of lights in the several connected parking lights were on. Only the two nearest the truck went out. I noticed that the lights were off for several weeks after the event. [14]

And later in this chapter we shall meet a New York SLIder

whose light *never* comes on again until the authorities fix it.

So much variety in the lights' on-and-off behavior defies our attempts to trace a pattern. An American living in Hampstead, North London, UK, was understandably perplexed by his experience:

> Two summers ago we regularly attended Saturday night performances at Kenwood, and would often walk home. On one particular occasion I was dumbfounded as I walked on Hampstead Lane, away from Kenwood towards Hampstead High Street. As we approached each streetlight it would go off, stay off for the duration of our walking past, and then go back on again. My recollection is that it would "flicker" on, but I can't remember for sure. [130]

If we could establish that there is one particular kind of light that comes on again more or less immediately, and another kind that stays off, it would take us a further step forward in our inquiry. But, naturally, most SLIders are not aware what kind of light they are affecting. Furthermore, it might not be a question of the *type* of light, but something to do with the way they are installed and controlled.

Alone or with others?

The SLIder isn't necessarily on her own, though this is more often than not the case. If she is driving with others, she or one of her passengers may draw attention to it. She may be met with disbelief, even mockery. But it's good to have someone to confirm that she's not imagining things, and if she's lucky another light will go off and her passenger will be shaken, if not totally convinced. Some typical reports:

> *A newspaper worker from California:* It often

happened when my boyfriend was with me, so at least he didn't think I was crazy. But it happens more often when I'm alone, driving, occasionally when I'm walking at night [23]

An American cab driver: The majority of the time I was alone, but occasionally SLI was observed when I was transporting passengers. [25]

Worcester, Massachusetts: It took a while to convince my wife, but since then she has been with me to witness the lights going out. [30]

We shouldn't assume that it is always the driver of the car who is the SLIder, though this seems to be invariably the case. So far as I know, there has not been a single instance of someone, who has done SLI elsewhere so knows she does it, also doing it while traveling as a passenger in someone else's car. Why not? The light surely doesn't know or care who's at the wheel.

How frequently do SLIders do it?

SLIders vary enormously in this respect. There are some who do it intermittently throughout their lives, so that it becomes commonplace, while for others it is a rare and notable event. We shall consider this again in later chapters when we look at the people who do SLI and their state of mind when they do it. Few do it on anything like a regular basis, though for some it comes pretty close. And there seems to be no pattern, for instance marking special types of occasion such as anniversaries. Typically, reports range from "once or twice a year," "twice a year," through "a few times a month" to "every month or so," "one or two per week," and even "'virtually every night."

Various SLIders report:

> At least 140 in the past year and a half.
>
> Typically, 3 lights will go off in an hour drive.
>
> Almost a nightly occurrence.
>
> Sometimes many will go out in one evening.
>
> I once blew out 7 lights in one weekend at a friend's house. His wife made me replace them!
>
> I would put out streetlights along an Interstate Highway exchange, there may have been as many as twenty lights involved at a time.

Some SLIders comment on the absence of any pattern. One reported "Sometimes it happens once a week, sometimes once a month, and sometimes even once a day." [31] Only a detailed investigation would establish whether this correlates with fluctuations in the SLIder's health, or with any other circumstance. Certainly nothing of the sort is conspicuous. Few maintain records, but even such figures as we have don't mean much as very few SLIders keep anything like an exact log. It is just something they are aware of experiencing a little or a lot, regularly or occasionally.

It's the same with the duration of time during which SLI is experienced. Some blithely say "all my life," while others narrow it down to a specific period in their lives. Interestingly, there are people who have patches of their life when it happens, alternating with long periods when it doesn't. The case of the young Dubliner cited at the beginning of the previous chapter seems to have been of this kind; he reports just two occasions, some time apart, when it happened to him. This likely relates to the states of mind that we consider in more detail in chapter six.

Although the greatest interest attaches to people who continually do SLI over the years, the once-only SLIders are a puzzle in their own right. An English astronomer, for example, told us it had only happened once in his lifetime, but that once was quite dramatic: "It happened late at night on my way home from a friend's house. During this journey 4 or possibly 5 consecutive streetlights went out as I walked past them." [146] He does not mention any circumstances that made him feel "special" that night. But there must be some explanation for why he did it just that one time.

Here is another "once-in-a-lifetime" story from a holidaymaker in the Canaries:

> Whilst on holiday in Tenerife last year, I liked to stroll on the marina last thing every night. As I passed each streetlight it went out. As I approached the next one, the previous one came on again. After three nights, it really spooked me. It happened every night for two weeks. It has not happened since. I mentioned it to my family, they only laughed and made a comment about faulty electrics in Spain. [77]

Meteorological conditions

Whatever the weather, SLI happens, but no pattern emerges in this respect. One informant tells us: "I can't remember ever popping a light in the rain, but then it hardly ever rains in Southern California." [13] A few SLIders report possible correlations; one from Cheltenham, England, tells us he did it when the weather was "cold and damp," [12] one from California, more surprisingly, mentions "cold and snow," [7] and another from Nevada "thundery weather." [25] A Swedish SLIder says "especially during cold periods" [204] but several report doing SLI from when they were on their holidays, strolling about in the evening when doubtless the weather was warm and dry.

The other possibility, that the lights themselves are affected by the weather, we have already ruled out; the purpose of the lights is to provide safety, and it would be ridiculous to design lights that could easily be prevented by the weather from doing their job just when they are likely needed most. Besides, if it was a question of the weather alone, every passer-by might be responsible.

How many lights?

Perhaps the most variable factor is the number of lights that SLIders affect, which ranges from a single one to double figures, sometimes here and there, sometimes in immediate sequence. It is rare for a SLIder to extinguish several at the same time, but one Californian [7] claims to have done this. In the previous chapter we showed that on technical grounds this is unlikely, but there could be occasions when a SLIder affects the controls that govern a number of connected lights. Most often, though, the lights are extinguished individually, and a comment from Augusta, Georgia, such as "some nights I hit the jackpot – 5 lights and more" [41] means one after another, rather than all at once.

Here are two particularly dramatic cases:

> *American cab driver:* I had been feeling uneasy all day.... At around 8pm I drove west along East 4th Street; along this street were sodium vapor lights spaced about 25 meters apart. As I passed, all of the streets on the south side of the street were going out, as I came up to a light, the next three would go out at the same time, then when I came to the fourth the next three would go out. [25]

No other SLIder has reported doing them in batches like this. Possibly they were arranged on a shared circuit.

Law enforcement officers are not often involved with SLI, but it can happen:

> *From a chemistry engineer in Woodville,*

> *Washington:* I was once stopped in a major city by police officers who wanted to know what I was doing to the city streetlights. They had observed that as I turned into a lighted street, each and every light went out as I reached within 3 meters of it. I looked back and nearly six lights in 1 ½ city blocks of streetlights were darkened! The police searched my entire car and me before allowing me to leave. As I drove away and on down the street, the oncoming streetlights extinguished just before my car reached each pole. I turned off the street and round the block and looked; the streetlights were still dark, and the police stopped me again and asked me *not* to drive down that street again until they had the lights checked out. I drove again up that street the opposite direction and once again the far side streetlights went dark as I approached all the way through the city approx. 8 kilometers. These were electric vapor type modern streetlights. [11]

Here is another intriguing variation:

> *Corona, California:* Sometimes lights that were out, in a row that was mostly lit, would go on. [14]

And a Texas woman adds to the confusion:

> It seems to happen only when I am very upset. As I drive under the lights on the freeway they either go out, or if they are out they come on. [76]

We have at least five similar cases where SLIders turn lights on without having first turned them off. And just to add a further complication, a SLIder from Los Angeles reports that "the light will not always extinguish completely; sometimes it will dim to the point of virtually no light." [34] Nobody else has reported this, but it may be that the dim glow is difficult to see. Or it could be true only of a certain kind of light.

Is it always the same light or different ones?

Some SLIders seem to have individual lights that they turn out more often than others. This is generally due to their location – they could be ones that their regular journeys require them to pass frequently, so this may not be significant. It is also likely, as we have already considered, that when some lights go out and others don't, they are the most fragile – older or weaker in some way. This may be what's happening here:

> *A television director from California:* At the time I was single, in an unhappy relationship with a woman who lived about an hour away, and without transportation, my car parked dead at the curb and requiring expensive repairs, I worked long hours out of compulsiveness and frustration. It was not a high point in my life. My walk to work took about 35 minutes, and in the darkness on the way home I was almost always thinking intensely. One night I was walking past an apartment building and the yard light turned off. After I had walked past it a few meters, it sputtered back on. I didn't think much of it until the next night, when the same thing happened. Now I began to wonder if perhaps I was doing this somehow, I made a game of it, actually concentrating on turning off the light – and found much to my surprise that I could do it consistently. I found that I couldn't turn off just any light, but that one was a snap. I discounted oncoming car headlights and even observed the

> light from a distance for a few minutes one night to make sure it wasn't just going on and off at random.

But before we jump to the conclusion that the effect was due to the characteristics of that one particular light, consider the sequel:

> Then one February night I was walking through the bitter cold and snow towards downtown Colorado Springs, feeling quite sorry for myself and actually getting angry. As my frustration with my situation rolled to the surface, the streetlights began to go out. Not just one at a time, but five or six at a time. As I walked down the street, more lights went out. I felt filled with excitement, as I looked back at the two city blocks of dim streetlights. [7]

Apart from the scale of the incident, this is one of the most vivid examples of a feature that we shall be treating in chapter six – the state of mind of the SLIder, which was manifestly the crucial factor in his experience.

Many other SLIders report targeting individual lights. For example, a student at Warwick University, England, reports:

> Just off campus there is a Tesco Supermarket to which access is gained through a dark pathway which is lit by streetlights. I noticed that on my way back from Tesco's, one particular light, and only one, would switch off as I passed it. The same light, and no other, was affected each time. [115]

Here again, we can reasonably suppose that this particular light was vulnerable in some way. But that does not explain why he alone caused it to extinguish when other people didn't.

It is not unknown for SLIders to strike up friendly relationships with particular lights. When I described SLI as a dialogue between the light and the SLIder, a woman from Alexandria, Virginia, replied:

> Hilary, it is interesting that you should use the term "dialogue" because I got to the place where I would talk to that particular light as I drove by (on those times when I was by myself). No actual conversation, just sort of daring it to happen while I was watching or acknowledging out loud that I had noticed the activity. [202]

Another instance comes from Canada: a bank official writes from Toronto:

> My wife and I noticed that certain streetlights would flicker and go out as we drove by them and/or for that matter walked underneath them. There was one particular streetlight that I dubbed my friend, and my wife and I would go out for our evening walk and I would go to it, touch it with my walking stick and talk to it. Needless to say anyone watching would think we had gone completely "bonkers," but nevertheless it would at that particular time go out and as we walked away I would turn round to say goodbye and it would flicker and come back on. There are also occasions when we were in the car and a car ahead of us would drive by and nothing would happen and then as we drove under a particular streetlight in the car sure enough the light would go out. [102]

In the next case, it is again evident that a particular light was involved, though in quite different circumstances. An office-worker from New York relates:

> I always put out the lights in the rest of the house when occupied with my photographic work. One night I noticed a small point of light in the living room. When I went to investigate, I found a narcissus blossom glowing, as it were. I checked for a source of light and found a thin beam coming through the shutters. I opened the shutters and saw the streetlight across the street. It immediately went out. An hour or so later the same thing happened. The light had gone back on; it went out again when I looked at it. [44]

The SLIder must have been a fair distance from the light – at least the width of the street, even if his house was actually at the edge of the roadway. This raises the question of the distance a SLIder has to be from the light, which we look at next.

The near and the far

The distance between the SLIder and the light is another surprisingly variable factor. Surprisingly, because you might well think this would be pretty much the same for everyone. Not a bit of it! Many SLIders do not specify an actual distance, but a correspondent from Wimbledon, England, says that he is about seven meters from the light when it goes out, switching back on again when he has passed. [167] We have seen that others find they need to be directly underneath it, and as soon as they move away it comes back on. At least five have reported this. A Hungarian engineer notes "These lights are 4-6 meters above our heads and when a light goes out, it is always the nearest one, not the one 3 or 4 poles away." [4] This is intriguing for, judging by what many SLIders report, there must often be times when a SLIder is within range of more than one. Yet it is always one alone, the closest, that he affects.

Yet the variations are extraordinary. Some simply say, "while approaching" the light, others have specified "2-3 meters distant," [12] "within 20 meters or so," [18] "about 25 meters," [35] "50 meters," [13] "while directly under the light or upwards of 50-100 meters away," [32] or even "up to 200 meters," [57] which is a really considerable distance! A SLIder from Harrogate, England, also calculated how far away he needed to be from the light before it came on again:

> I would usually be over 100m before it would revert to its original state. On the approach it would vary from 5m to 10m. [178]

As though to contradict the SLIder who needs to be directly underneath the light, another from Aston Clinton, England, specifies the opposite:

> The light invariably tends to blink out when I am about 15-20 meters away – it rarely happens when I walk under them. [199]

To account for this huge variety in the distance over which the SLI-force operates defies conjecture. Are some lights fussier than others? Are some near-sighted? We might imagine that local conditions could play a part – perhaps the force is stronger on a rainy night; sultry or clammy weather may make a difference. Or is it a feature of this particular SLIder's SLI-force?

We are taught that any force diminishes with distance, and we would expect SLI to do just that. Certainly, distance is evidently a factor in those cases where a sequence of lights go out as the SLIder approaches, come on again when she has passed. But if five or seven meters is a typical distance, how come some SLIders report much greater distances? How could the SLI-force be strong enough to extinguish a light 50 meters away? We might be tempted to think that the more distant ones are due to coincidence or misinterpretation of a defective light, but we do not have the right to ignore any witness statement, however improbable. Later in this chapter we quote a medical practitioner who went to some trouble to observe his SLI powers, and found that they were operative over an astonishing 300 meters – which we would find hard to believe if this particular SLIder hadn't been so painstaking to establish the facts.

Damage

Damage is rare. The lights generally come on again as if nothing has been done to them. But there are exceptions. An informant from Bournemouth, England, seems to have been unusually destructive:

> One time, after having an argument with my parents, I did seven lights – three of the light bulbs smashed when they went out. I have occasionally done lights on command, and have damaged a couple of hundred lights in the last 5 years. [158]

Damage can occur with domestic lights too. A woman from Leeds, England, reports:

> [After describing other dramatic incidents:] Someone else I was very close to moved out of the house. I was very sad, and then as I was sitting on the sofa breastfeeding my baby, the angle poise light exploded showering us both with glass. [170]

And an office worker from New York reports:

> One light in particular is near my house and I am turning that one off more than any other. It is then off for about a week until the highway department fixes it. Then I put it out again. [44]

We have already suggested that when SLI affects some lights but not others, this may be because some are weaker than the rest. But intriguingly, the SLI effect can be operative even when it is clear that the light is already malfunctioning. A SLIder from London, England, reports, after describing other instances of causing SLI:

> Recently, one streetlight in Sandringham Road where I live did begin to fail. One could see it from a distance flickering and it would go out if I walked past. I began to pay attention to see if other people walking by it had a similar effect. Sandringham Road being quite busy, whereas the previous lights had been on much quieter roads. I have to admit that the light did not seem to be particularly selective about who it went out for. It was by no means popping on and off every time someone walked by, but I was certainly not the only person to apparently affect it. One person did walk past it and out it went, then 30 seconds later I walked by and back on it went! [176]

This suggests that the light was a particularly weak one, perhaps in its death throes. It seems that quite a number of the passers-by affected the light, but yet not all.

SLIding at will

The first SLI incident is invariably spontaneous. But after that some SLIders become curious and start to experiment, either because they want to know if they are really doing it, or to try to find out what is going on, or simply to impress their friends. Some examples:

> *A housewife from North Carolina:* Usually, the lights go out by themselves, but occasionally I'll look at one, try to make it go out, and it does. [1]

> *London, England:* On one occasion when I was with a girlfriend, I predicted that the light – one that I regularly extinguished – would go off as we approached. Sure enough, to her astonishment, it did. [167]

> *A man from Fair Oaks, California, tells us of* ...a neighbor's yard light, which I could induce to turn off by walking past it and concentrating. (I did so once for the benefit of my mother-in-law, who was visiting from Colorado. As a devout Catholic she wasn't exactly thrilled that I could do this.) I did verify that the light was not equipped with a motion detector on/off switch. My wife could not induce the light to switch off by itself as she walked past it. [7]

> *Woman from Carlsbad, California:* I did it once *very* consciously for my husband. [27]

> *Woman from Athens, Greece:* We were dropping friends off at their apartment building on one of Athens' hills; as I stepped out of our car the light opposite popped and went out. As a joke, my husband, who was used to this by now, said to me, "Do your spookies – make it come on again."

I lightheartedly took up the challenge – I clicked my fingers in the direction of the light and it came back on again. At this, the very narrowly brought up Catholic lady who was with us fled for the cover of her home. She told me the light subsequently went out again and never came back on. [61]

Rohnert Park, California: I decided to see if I could turn it on after I would walk under it and it would go out. I would be about 40 meters past it and stare intently at it wanting it to come on and it would – sometimes. [35]

Teacher from East Wanatchee, Washington: On several occasions I have caused to "will" streetlights to go off, and it's become easier over the years to do this, though by no means completely reliable. One time, I caused a streetlight to arc and emit a shower of sparks, simply by saying "HA!" at it, when it had already gone out. It was a bit frightening, to say the least, another time, one went out, and I joked to my son, "Watch me put out the rest," and by golly, I did! [86]

Gainesville, California: While waiting for a bus with nothing to do, seeing a streetlight and trying to get it to go out, amazingly it did, after a few seconds it came on again and so I concentrated and off it went again. That's when I noticed this apparent trick. [45]

The greatest degree of control reported to us came from a doctor of chiropractics, accustomed to working with healing energies:

> Being analytical on both a scientific and esoteric basis as well, I found that I could turn these lights off at will or not turn them off. I realized that I could do so at a distance as much as 300 meters away. After some experience of attempting to perceive what exactly I did to make this occur, I found that I could reverse this energy and turn the lights back on. I can do this, turning lights off and then back on, over and over, even in rapid succession at times. [57]

Another SLIder tries to explain her answer to the question, Have you ever made a conscious effort at SLI?

> *Cunard, California:* Yes – but this is a subtle, oblique sort of effort. I will be aware of myself having the thought on another level of thinking, and then a few seconds later the light goes off. [46]

This raises a very interesting question, which unfortunately we are not yet in a position to answer, which is: even if the SLIder is unaware of doing SLI until it happens and is taken by surprise when it does, is there in fact a hidden part of her – operating on a different level – which *does* know this? We shall touch on this again in our final chapter.

If we could control SLI at will, investigation would be much easier. But time and time again, SLIders will tell you that when they try to do it consciously and deliberately, they fail, but then, they add, after they have given up trying, it will happen spontaneously. Here is an example from a professional magician who told us:

> I remember walking under them trying to make

> them go out but I couldn't. The moment I stopped willing it to happen it would start again – like someone catching me out. [132]

This is a very important feature, possibly the most revealing feature of all reported SLI behavior. It tells us, beyond any doubt, that the SLI effect is emphatically not an automatic effect, for there is surely no possible way in which the light can know whether the SLIder is or isn't consciously willing it to happen. The will of the SLIder is, beyond question, the operative factor.

But it's not a reliable factor. The occasional instances when SLIders *can* do it to order tantalize us. But in the overwhelming majority of cases, SLI happens only when the SLIder is not consciously exercising her will. So when we say it's her will that is the operative factor, we are not talking about a consciously directed exercise of will power. Just what we are talking about is buried beneath the surface in her unconscious or subconscious.

Why should this be? The key can lie only in the personality of the individual SLIder. It's good to know the design and modus operandi of the lamps because that tells us what the SLIder is actually *doing*. But they are only contributory factors. We shall find the secret of SLI only by delving into the mind of the SLIder. We will try to evaluate this crucial feature in our final chapter.

Foreknowledge

Several SLIders sense that a SLI event is about to occur, but it is not a very specific premonition – just a general feeling. A teacher from St. Paul, Minnesota, tells us she had her first SLI when she was 14, while out with her mother. She says:

> It always happens spontaneously but I can "feel" often now where and when it will happen. I can walk down a street and tell you "Watch that light!" and I am often right. Sometimes I know it with almost 100% assurance that it will occur.

> … I definitely cannot tell what is going to happen in the future. I personally feel just sensitive to energy in general and I just kind of have a sense if a streetlight might go out, or if there is an intense energy in a room. I think it's just from learning and experience. I can tell if the light will flicker. [211]

Others who have mentioned this kind of feeling do so in much the same terms – not exactly foreknowledge, just a sense that something is about to happen. Since the light cannot possibly know that a SLIder is about to come along the street, and the SLIder cannot possibly know that she is about to meet a light that she will extinguish, this is one further impossibility to add to the whole crazy scenario.

Chapter Five

Tinker, Tailor, Soldier, SLIder

> I am a perfectly ordinary unimaginative down-to-earth Capricorn person, a pensioner to boot – and after a lifetime of putting out lights it still takes me by surprise every time it happens. I look forward to your research. Yesterday my neighbor was complaining about getting electric shocks from her car. I said, "Oh, that's ok – it's when you put out streetlights you have to worry." [131]

There are no professional SLIders. Until we learn how to harness this wayward skill, it is not a career option. This SLIder from Greenwich, England, adds:

> I don't put out *every* light *every* night, but I have often put out the very large lights on the Dover Road as well as ordinary streetlights. It amazes me – how can it possibly happen? [131]

How, indeed? Like her, the hundreds of SLIders who have told us of their experiences are "ordinary unimaginative" people in every other respect. You wouldn't know a SLIder if you met one (and you probably have, without knowing it). They come from every walk of life, every age group, and as far as we know, from every country on the planet. What is it that makes them act differently when it comes to streetlights?

A famous poem by Walt Whitman is headed "I sing the body electric."[5] The two SLIders just cited, like the rest of us, are electrical machines; we need internal electric power to perform every action, think every thought, and simply to stay alive. Whatever SLI is, we can be sure that electricity plays some part in it. It is tempting to think that what makes SLIders special is simply that they have got more of the stuff than the rest of us – that they are, whether chronically or temporarily, "supercharged" – many SLIders actually use this or a similar term – and that they need to release this excess, and that SLI is the way they do it.

Obviously that's absurd, as it stands, but there may be a grain of truth in it. It makes sense to start from the position that SLIders are people whose natural electricity is being directed or diverted in an unusual way. Because SLIders are so like the rest of us in every other respect, it could well be that they are making an exceptional use of something we all possess.

A fundamental question is: Does SLI just happen to people, or do people make SLI happen? In the light of what we have already seen, the answer is, unquestionably, the second. SLI doesn't happen to everyone who chances along; it happens only to certain people – or perhaps, to certain *kinds* of people. But we could be mistaken. An alternative possibility is that the SLI-force is something we *all* potentially possess, you and I included, as part of our biological make-up, but perhaps it manifests only feebly most of the time, so that we never notice. If this is the case, SLIders are simply people in whom the force manifests more strongly, and so more noticeably. We must bear both these possibilities in mind.

And another thing: Could SLI be happening more frequently than we suppose? After all, lights blow for all of us, only we customarily mutter resignedly to ourselves that sooner or later it was bound to happen, and just growl a little because it happened while we were hunting for our kid's toboggan in the attic. But suppose that, in fact, some or even many of these commonplace extinctions are the consequence of SLI?

This raises the intriguing – but alas, unanswerable – question: Have there always been SLIders? If it is a natural force emanating from the individual, were there SLIders in Ancient Rome? (They would have needed something more sophisticated than flaming torches to demonstrate their skills on.) Or we can wonder if there are nomads in the Sahara Desert who, if they were brought into Fez or Timbuktu, would do SLI like their urban cousins? Are SLIders the product of the urban environment? Was the faculty dormant until street lighting was installed? We shall return to these and other far-out speculations in our final chapters.

SLIders rarely have the chance to check that they're doing something that other people can't or don't, but this informant from Normal, Illinois, had the opportunity to compare his performance with that of others:

> SLI has been happening to me for years. Recently my mother and I stopped at a highway rest stop just north of Springfield, IL, to stretch our legs after many hours of driving. The previous day I had showed my mother that I made streetlights turn off and she thought this was really odd.
>
> While stopped at the rest stop, we both walked towards the restrooms and the light above us went out. This was the perfect opportunity to prove it to her. When we had come out of the restrooms, the light was back on again. So I walked toward the light and it went out again. To what extent, if at all, is SLI the consequence of living in an environment where there are streetlights?
>
> We spent about 26 minutes individually walking toward the light pole and it went out every time on both of us. We stayed around and watched other people walk under it, but nobody else had an effect on it. [26]

There seems to be an implication here that his mother, too, was affecting the lights. This is one of very few cases known to us that suggest that SLI might run in families, but there *are* others, as we shall see later in this chapter, and they add a new slant to the question: Why to some and not others?

What kind of people do SLI?

> *A woman from Valencia, California:* I thought it strange (or interesting) that a man from whom I was purchasing vitamins asked me "out of the blue," what I thought of an experience he had been having. He said that as he would drive along, streetlights went out one after the other, what did I think it meant? I thought it strange he should ask me such a question. I was a stranger to him. [46]

Does it take one to know one? Was this little incident pure chance, or did the salesman sense some kind of affinity? Was she unwittingly sending out "signals" that she, too, was a SLIder? Certainly there are no discernible outward characteristics. Not every SLIder tells us about his/her occupation, but it is clear from those who do, that they may come from pretty well every trade, profession, or activity in the Yellow Pages. Here are just some of those represented in this book: accountant, Air Force officer, alarm engineer, artist, astronomer, automobile technician, chemistry engineer, computer engineer, counselor, dermatologist, draughtsman, editor, electrician, geologist, hospital worker, housewife, lawyer and law student, mathematician, psychic medium, microbiologist, minister of religion, musician, nursing student, psychologist, psychotherapist, radio announcer, singer, software engineer, teacher, television art director... In short, you could hardly find a more varied collection. If only they were all vegetarians, or born under the sign of Capricorn. But no such prevailing factor emerges from their histories.

Pretty well all SLIders display curiosity. One of our earliest informants, writing from Oregon, made this comment:

> It occurred to me that the ones I zap are all on light-sensing switches, and perhaps my energy at certain times for who knows what or why, is the right kind and strength to trick the switch into thinking it is daytime. (2)

It is important that SLIders should be inquisitive about their experience, for there can be little doubt that SLI is by nature both objective (as regards what they do) and subjective (as regards what they feel about doing it), and the SLIder's personal feelings can give us insights as valuable as any offered by engineers.

However, there is more to people than how they make their living, and we could speculate that there is some pattern of emotionalism, some personality trait that they all share. Short of drawing up a psychological profile of every one who contacts us, we can only draw our conclusions from what they say about themselves and how they express their feelings. Reading their letters and emails, one of the first things that strikes us is that very few SLIders are worried or anxious about doing SLI. At the same time they are always pleased to learn that they are not alone, and they are delighted to find that what is happening to them is not a symptom of mental illness but a scientific anomaly that is taken seriously and investigated. Here is a typical comment:

> *Harrison, Arkansas:* I cannot tell you how happy I am to read about other people who share my experiences. I thought I was the only one who ever had problems with streetlights going out. I first heard about it in an article in *OMNI* magazine and almost dropped the magazine when I came across it. [89]

Sentiments to much the same effect are expressed in nearly every reply we have received. By and large, we would have to say that a more level-headed, balanced, common-sensible collection of people would be hard to find. They are not fringe folk, kooks, or nuts, far from it. Without exception, they write to us in an open and rational manner, in which puzzlement and curiosity are the primary feelings, coupled with a light-hearted approach. It is a real pleasure to read their letters and emails. And it is good to know that it is a reassurance to them to learn that serious scientific interest is being directed at their stories. Many people who experience SLI are concerned in case it is a sign that they are somehow "odd." They are invariably relieved to learn they are one of many. Two typical comments:

> *Taylor, Michigan:* I can't tell you how much relief I have felt since hearing others have experienced this sort of happening in their lives also. [5]

> *Monument, Colorado:* I had this problem but had no idea it was widespread enough to be a *phenomenon.* [13]

Similar comments are made by virtually all SLIders when they learn that the phenomenon is widespread and that it is not an indication that they are "seeing things" or "going soft in the head."

How old do you have to be to do SLI?

A correspondent from Norway reveals that boys will be boys in his country, as everywhere – and that SLI happens to them as it does to adults:

> I think I was about ten years old the first time it happened or the first time I noticed. The first time I knew that this could not be by accident was when we were "playing" as kids, kicking the light poles so that the light would go out, but very often

> the light would go out before I got the chance to kick the light pole. Now it happens maybe two or three times a week. [214]

The SLIder's age does not seem to be a factor, especially since many SLIders report doing it over a period of years – as long as they can recall, in some cases. A Swedish SLIder tells us he has done it "throughout my entire life, though much more frequent during my childhood." [204] Several SLIders are retired persons; they tell us that it happens when they are out walking the dog. But a great many do it in the course of their working life, often while driving to or from work.

Adolescents and college students seem to figure disproportionately among SLI reporters, and there may be an interesting reason for this. As we shall see in the next chapter, the SLIder's state of mind is often a significant factor, and teenagers and young adults are particularly liable to growing pains and emotional hang-ups that seem to favor the manifestation of SLI.

Do young children do it? We have yet to hear of an infant putting out lights as his baby carriage passes beneath a light, but a 17-year-old student from Oregon, though he can't exactly remember when he first noticed it, recalls:

> The first one that I can remember was at a local supermarket when I was about nine years old. My mother and I were approaching a streetlight when it just blinked out. It stayed out until we were probably 60-90 meters from it. [181]

He mentions that at about the time of his first sighting, he had an accident while swimming when he grabbed hold of an electric cable, which gave him a severe shock, though with no serious effects. But there could be a connection, in every sense of the word.

> *Brooklyn, NY:* I first noticed that streetlights went

> off in my presence when I was about 14. I'm 32 now and it still happens. [20]

What about gender?

Analyzing the first two hundred reports we received, it emerged that 140 were from men, 60 were from women. However, I don't think too much should be read into this, as other factors may intervene. For instance, a good number of the early reports came in as a response to an item written by Dennis Stacy in the popular science magazine *Omni*, which may well have been read more by men than women. On balance, it is safe to say that SLI is practiced about equally between men and women.

How reliable are SLIders?

We have not the slightest doubt that the people who contact us are telling us of things that really happened to them. Of course sometimes they may be mistaken, they may be misinterpreting what they have experienced. But that does not make them any the less reliable as observers; it merely warns us to be on our guard against jumping too readily to conclusions.

But I don't think there is much danger of that, given the diversity of the reports and the elusiveness of explanations. Because SLI is almost unknown to the general public, there is little danger of their testimony being distorted – however sincerely – by their expectation of what they *ought* to have experienced. There isn't a model for SLI that is widely known. Other anomalous experiences, such as seeing a UFO or a vision of the Virgin Mary, are liable to be affected by what the witness has heard of previous events of the kind, which is why reports of such experiences tend to run to a pattern. So far as the SLIder is concerned, his experience is one of a kind. SLIders have no expectations. What they say, therefore, will certainly be honestly reported.

Though, as you see, our informants come from many parts of the world, we have met several of them personally and discussed their experiences face to face. There has never been the least suspicion that any of them were making up their stories, or misreporting their experiences.

Is SLI related to ill health?

Virtually no SLIders tie their ability to any form of illness. However, there are exceptions, and they are interesting because they introduce a secondary element into the equation of person and light. An informant from Aylesbury, England, reports:

> In October last year I suffered from a recurrence of the occasional attacks of Hypomania I am inflicted with. One morning as I walked through the station car park in Wendover the lights came on as I walked under them. Eventually they were all on. They are sodium type lights on very tall poles. It was a most disconcerting experience. Initially I thought that the first ones had come on because it was a fairly dull morning, I then realized that my presence was probably the trigger. [109]

This informant also stated that while in a hypomanic state he causes interference with the hi-fi. In chapter 8 we shall meet a SLIder with SLI and other powers, who attributes them to illnesses earlier in her life. [206]

An American SLIder moved from urban Southern California to rural Colorado, but reports that "I still keep zapping them though not so frequently." In her thoughtful letter she wonders if her SLI potential is connected to her illness:

> Last year I was diagnosed with Multiple Sclerosis. I believe that the SLI phenomenon can be traced to naturally occurring or disease-produced interruptions in the conductivity of neural synapses. In MS the neural sheaths are literally

> eaten away and replaced with poorly conductive plaque. If conductivity problems are the culprit, it should be possible to detect this with sufficiently sensitive sensors and appropriate stimulation of the affected neural pathways.
>
> Even though the explanation seems sensible to me, it still doesn't explain exactly how it works to produce SLI. I think it unlikely that conductivity problems in the brain could translate into the type of voltage increases required to affect SLI at a distance. Perhaps we don't know all there is to know about electrodynamic processes? [13]

Indeed, we surely don't. This lady rightly admits that she can't imagine how the SLIder's condition translates into the SLI process, but of course that is the fundamental problem with SLI anyway, whether the person who does it is sick or healthy. It's a line of thought that is certainly worth following up.

Quite a few SLIders mention having a sensitivity to everything electrical, but only as a possible side-effect. We had an interesting letter from an Australian teacher, who, though she does not do SLI as such, thinks that her condition may be related:

> While I don't have this specific problem, I suffer from electro-magnetic radiation sensitivity, which I think may be related. With the advent of the mobile phone industry I have become increasingly sensitive to all EMR [Electro-Magnetic Radiation] devices – TV, radio, computers, phones and mobile phones. I suffer from unpleasant physical sensations when exposed to EMR, followed by exhaustion, mental fatigue and disorientation. Sleeplessness is the worst and most ghastly

> aspect of it. My theory which occurred to me when I read a newspaper article about SLI is that these much higher levels of EMR in cities is having an effect in the electro-magnetic field of every individual. Perhaps healthy people are able to assimilate the energy into their own fields, thus affecting electrical events round them when they are disturbed. This would fit with the sexual aspect mentioned in the article, which is linked closely to human electro-magnetic activity. [108]

Electro-magnetic sensitivity

The account just quoted mentions electro-magnetic sensitivity in connection with SLI, and some interest has been shown in SLI by various organizations that are investigating, or concerned about, sensitivity to electric, electro-magnetic, and electronic appliances, raising the possibility that there is a connection. Dr. Jean Munro, consultant to the Breakspear Hospital at Hemel Hempstead, England, has pioneered research into the dangers arising from exposure to EMR, and into beneficial treatments.[6] A few other SLIders have mentioned EMR as possibly being associated with SLI, and some mention their own sensitivity to such devices. But most do not, and it would be difficult to establish a case for any kind of overall correlation: Why should the one have anything to do with the other? No SLIder has ever reported any effects to their health from doing SLI; it appears to have no medical significance, whether positive or negative. If anything, SLIders are likely to report a sense of exhilaration after knocking out a light. So, while this is clearly an area where we need to keep on the watch, right now it does not seem a promising line of research.

Does SLI run in families?

There are occasional indications that close relatives share the ability to do SLI. A letter from London, UK, reports:

> My younger sister and I both used to be able to do this regularly. At the time (I must have been in my early twenties) my sister and I would treat it as no more than a peculiar "party trick" with which to amuse (or frighten!) friends. [101]

And an informant from California mentions that three or four members of her family seem to share the ability:

> Regarding Electric People, my mother, myself, and my brother have been putting out streetlights for as long as I can remember. My mother, sister and myself can remove, collect and discharge static electricity from clothes, hair etc., but never suffer or be bothered by it. I also have trouble with radios and stereo equipment, tuning them. [155]

An informant from Exeter, England, writes:

> I have talked with several people of this effect and curiously enough, the only other person who seems to have experienced something similar is my father, who also notices that some streetlights are more susceptible than others. [195]

Two sisters from California tell us they both do SLI. [72] A report from Fort Myers, Florida, involves a mother and son. [47] And we also have a husband and wife from New York who both do it. [74]

Is there a PSI-SLI connection?

Perhaps because streetlights are such mundane, matter-of-fact things, no SLIder seems to think that anything supernatural is happening when they go off or on in her vicinity. If she did so, she could be forgiven, since what she does seems to be happening outside the laws of science as we know them, and many SLIders associate what they do with other happenings on the fringes of scientific knowledge. But just because you can't explain something, doesn't make it supernatural. Like other oddities, it is something that happens to them without their choosing to do it. Consequently many SLIders tell us about other anomalous experiences. These include:

Seeing apparitions [11] *and seeing auras* [47]
Doing automatic writing [47]
Having deja vu experiences [54]
Doing healing [6, 37, 61]
Having intuitions [37]
Experiencing precognition/premonition [3, 11, 13, 47, 48, 54]
Experiencing telepathy [52]

One mentioned "and also I've shocked boyfriends while kissing them." [3] A SLIder who says she is "very into New Age ideas" says that she is *very* interested in ley lines:

> I had an awareness of ley lines since childhood. Probably have a transmitter inside nose, placed there by aliens. [45]

(We don't think she is being serious.)

A woman from Hawaii reports:

> I have also worked with biofeedback equipment and am able to make the dials and temperature gauges fluctuate with ease. I was asked to demonstrate the equipment at the University of Hawaii. Is there any connection? [58]

A student from Represa, California, relates an incident in which he was having a tense argument with his girlfriend and a roommate, when a loud bang occurred that made all three think one of the others had fired a gun, but they could identify no evident cause. [59]

This is probably pretty much what we would expect from a cross-section of the general public, and certainly doesn't imply any direct connection. However, a 24-year-old musician from Melbourne, Australia, could be an exception:

> I haven't experienced SLI as such but I have experienced related phenomena on and off for years now, of blowing light bulbs fuses and unwittingly damaging stereos. It became more pronounced this January which was a time of great stress as a very close relative slowly died. In one week I broke three tape-recorders by pressing "Play" on them. One of the breakages was purely mechanical, another was a fuse and a third was a brand new stereo that fused when I pressed a button after unwrapping it fresh from the factory packaging. I also fused three light bulbs that week.
>
> I have also lived in a house previously which had, to our knowledge, never had any electrical idiosyncrasies with the wiring, fuses etc. But after living there for 6-7 months the whole house was fusing constantly and light bulbs blowing when I switched them on. Note that this house was extremely noisy from traffic to the point where my menstrual cycle became random and unpredictable, indicating a very fundamental physical disease.
>
> While my close relative was dying I went to see a naturopath about an intense stomach ache, and she asked me, totally unprompted, if I was blowing fuses a lot, and that my belly ache and electrical interference was an obvious

expression of this grief. Prior to this I have never believed in naturopathy. But am pretty converted as a result.

It might help to know that I am a clairvoyant part-time which funnily enough is a line of work I accidentally fell into after taking advice from a clairvoyant in my quest to stop blowing light bulbs. (I'm actually quite cynical about any kind of psychic stuff and am remarkably down-to-earth about my line of work, but after being recommended to try doing tarot cards to stop blowing light bulbs, I found it very successful.) I sought help from this clairvoyant due to pressure from friends who were getting spooked by my "talent." I was too embarrassed to consult a doctor or any related professional. I found tarot reading very successful and I was surprised to establish a base of return clientele over time. I've always had predictive dreams but never thought anything of it. Working as a clairvoyant and taking mineral supplements (I have a tendency to deplete minerals such as magnesium and sulphur) pretty well cures me.

The only other thing I can think of is kinda random, but I've had stacks of x-rays in my life and had electric shocks, 3 or 4 times. I also get static-electric shocks from certain kinds of carpets whether I'm wearing shoes or not. [156]

This fascinating letter contains a number of lines for further inquiry, which might tell us something about the chemical make-up of SLIders and other electric people.

More typical is this comment by a French SLIder:

I am a rather famously cynical person believing

> in virtually nothing – God or gods, astrology or parapsychology, or even the idea of anything after death. It wasn't likely that I'd be searching after the causes for this truly wacky phenomenon... When a friend and I were walking one dark October night in Prague and the lights, one by one, went off as I passed and then came on again we got wildly giggly but that was all. I have noticed the same thing happening here in France, having retired from an American university where I ran arts programs for years. [133]

Overall, we can rule out interest in the paranormal as a major factor, if only because so few SLIders even hint at such an association. But there are a few who do suggest that there may be a psi-connection; like those who report a possible connection with sickness, they open up lines of inquiry that could be potentially helpful. But as things stand, it seems more likely to be the other way around. Rather than looking to the paranormal to explain SLI, perhaps we should look to SLI to account for some of the weird incidents we label paranormal.

Chapter Six

Alternative states of mind

> *A student at Washington University:* There had been an argument, and Dan was furious and a bit embarrassed when he left the dorm one night, storming off toward campus. I was actually worried that he might hurt himself and so I followed him, staying within eyesight but not so close as to intrude upon his space. He knew I was there. He stomped around campus, and it was late enough that the two of us were pretty much the only people around. Eventually I noticed that every time Dan approached a streetlight, it would go out. Then when he had passed, about the time I reached the light, it would come back on. Dan noticed too, and he stopped under one that had just gone out, looked back at me standing under one that had just come back on, and then came toward me. As he walked to me, the light over my head went out and the one he had left went on again. He got curious, and started to test it, but after a few minutes of walking the effect no longer happened. The first two or three we approached went out, then no more after that. Maybe his black mood turned off the lights… [210]

We could hardly wish for a clearer account of the SLI process, and all the better for being reported by a non-SLIder who followed in his tracks but manifestly didn't have the same effect on the lights, though after their argument we might have expected both of them to have been in the same state of mind. This is an exceptionally unambiguous example of what is clearly a crucial feature of SLI:

the feelings of the SLIder at the time. It takes many different forms. Here is an example from Leeds, England:

> My experience is that when my partner left me, I was devastated and really sad. Within a few days of him leaving I had to replace virtually every light bulb in the house as they all went. [170]

These SLIders are telling a story that is repeated frequently in SLI narratives – bereavement, loss, anger, anxiety, frustration, and depression are repeatedly associated with manifestations of SLI. But why? What has the inoffensive, unoffending streetlight done to become the SLIder's victim? Okay, so the SLIder is feeling pent-up emotionally – but why take it out on the light? Imagine what it would be like if, every time you wanted to switch the kitchen light, it had to measure your mood before deciding whether or not to do what you wanted!

Besides, the emotions are by no means invariably doom and gloom. Elation and exhilaration, joy and jubilation, can also trigger SLI. Though lows are the more frequently mentioned, highs figure largely also. It seems to be the *intensity* of the emotion, not its nature, which sparks off the SLI response. Whatever the feeling, there's got to be lots of it.

Are SLIders people who take things to heart more than the average? Are they people who respond more fiercely to emotional situations? Such possibilities may seem to be strikingly demonstrated in the experience of an Australian factory worker:

> I was at an extremely low point in my life, having recently separated from my wife and being left with considerable debt. I was working in a miserable metalworking factory on ten to twelve hour afternoon and night shifts six and seven nights a week. As a consequence I regularly found myself driving the 40-odd kilometers

from work at Moorabbin to where I was staying in Mount Eliza between midnight and 6am. As I would drive along the Nepean Highway I began to notice that streetlights would go out just as I drove past them. Not every light, but on some nights maybe up to a dozen.

At first I thought it was just coincidence, but simple calculation suggested that if the lights went out at this rate for everyone there would be no operational lights at all. I asked friends, relatives and co-workers who used the highway at night if they had noticed a similar phenomenon and all told me they had not.

During the time I noticed the phenomenon I was very depressed, almost suicidal, and it became apparent that the lights went out when I was deeply, even morbidly introspective. When it began to happen more often it sometimes made me paranoid and even to question my grip on reality. More often, though, I found it would break into my self-obsession and get me to thinking not only about the light business but also about wider philosophical, metaphysical and existential issues. I never saw a light go out while I was looking for it to happen. However, if I drifted back into unhappy introspection, it would often happen again.

Since my life circumstances began to brighten I have not noticed anything like the events I have related happen again. [112]

Most of us have noticed that when we are angry, upset, or otherwise stressed, we tend to drop things or get flustered when performing simple tasks like lacing up our shoes. We have already seen many cases where the SLI events take place when the SLIder is in an unusual state of mind, sometimes recognized at the time, sometimes realized only in retrospect. This rarely amounts to an Altered State of Consciousness (ASC), such as dissociation or loss of consciousness altogether, but often SLIders report being unusually depressed/excited/impatient/angry but, paradoxically, they can equally be also exceptionally calm or relaxed, or happily excited. The following report is from a SLIder from Taylor, Michigan, who had no doubt that his SLI experiences and his state of mind were closely related:

> My experiences started around 17.9.1987. On this day I had what I can only explain as "a spiritual awakening to my oneness with all in the universe." For 7 straight days I experienced a "heightened consciousness." Not sporadic, but for 7 straight days. During this time, while driving my much familiar road home from work, streetlights would go out as my car would come to a stop at intersections. At first I didn't think so much of it, but it began to be very frequent, so I began to pay closer attention to what I was feeling or thinking when these occurrences took place. My initial reaction was, "Hey, who do you think you are to be able to cause streetlights to go out?" but again, it became more than mere coincidence.
>
> When these incidents take place, I am usually feeling as though spiritually or emotionally reaching out in all directions and at peace, also, my mind is usually left to go "just where it would," as you might say. [5]

It's surely an important clue, the way SLI experiences relate to state of mind, but it's a baffling clue, that it should vary from one emotional extreme to the other. No pattern, no common feature. It isn't even the case that an individual has to be always one thing or the other. A SLIder from Pewaukee, Wisconsin, reports: "it seems to be tied to strong emotions *(positive or negative)*" [36] (his words, emphasis added). Later in this chapter we shall read another SLIder's statement that he, too, does it in either frame of mind.

A computer graphics instructor from Darlington, England, reported a particularly striking occasion:

> It started with one particular light outside a local printer's and ended with me arriving home at night in tears and very distressed as three or four lights went out as I passed under them. At the time I was suffering from a quite severe depressive illness brought about in part by the deaths of four close relatives/friends in a short space of time... After several times the coincidence became noticeable and I mentioned it to my wife. It all seemed so pointless, but then she reminded me that the night my father died a night light in the house started to flicker on and off for no reason. I thought for a while and then replied that if he was going to choose any light, he'd obviously choose one above a printer's – he was a printer, and most of my family are involved in print and graphics. [140]

Among the states of mind specifically mentioned in so many words by SLIders are:

Feeling agitated, aggravated, upset [19, 56, 60]: "When I get aggravated, fluorescent lights go out over my head, or over the heads of those with whom I am upset." [1]

Feeling angry [3, 20, 37, 55]: "irritated after crowded subway journey" [9]; irritated when waiting for fiancée to show up [78];

repressing anger [17]

Feeling anxious, uneasy [25, 37]

Feeling tense [37, 46]

Feeling worried [14, 37]

Feeling stressed [6, 52]

A woman from Minneapolis goes into detail:

> The kind of stress that is prolonged and not acute at any particular point, and which for that reason must be submerged and lived with, in order to continue daily functioning on the job, etc. The kind of anticipatory stress involved with waiting with a loved one who is in the death process over a period of a year or so, for example, or the kind of psychic turmoil which an individual undergoes over a long period of time, when undergoing a transformation in values, understanding and development. [6]

Concentrating [4, 13, 22, 50]: A Hungarian engineer was in a strange town where he had been called in to resolve a difficult problem that absorbed his mind as he walked back to his hotel. [4] By contrast, though, a Vancouver resident tells us he sometimes has SLI happen "after I've come to some sort of solution, resolution, or some measure of peace about a problem." [207] No other SLIder has reported anything like this, and it is rather puzzling, for whereas concentrating on a problem might well be expected to cause stress, when the problem is resolved you'd think the individual would relax and lose his stress.

Distracted: From Aston Clinton, England: "They most often blink out when I am in a distracted or troubled state of mind. At these times, I tend to try and walk it off and talk to myself as I walk. As I do this I'll be quite absorbed and I'll look up and a streetlight will blink out." [199]

Feeling unhappy [3, 7]; feeling emotionally shut down [16]:

feeling fed-up [48]; feeling depressed [12, 47, 48, 55].

Feeling happy [20]; feeling elated [47, 55]. A North Carolina housewife tells us "SLI mostly happens when I'm elated, but being upset causes it as well." [1] An informant from Glen Cove, NY, says: "It usually occurs when I am in an extremely good mood, driving in my car singing to the radio or trying to get somewhere in excitement." [151]

Feeling excited [48, 50, 56]; "high energy level" [27].

Intensity of feeling [6,7]: "This seems related to intensity of feeling that is for some reason not being expressed, or expressed *fully*." [6]

Feeling irritated [52] but note: "I do notice that I need to be feeling more irritation for household lights and appliances to malfunction but streetlights go off around me without me being really annoyed." [9]

Feeling relaxed [40]; feeling serene [5]; "nothing on my mind." [60]

Sexual activity [38, 54]: A psychologist from Great Falls, Montana, remembers:

> Many years ago when I was dating a particular girlfriend this phenomenon took on a more evident effect. This girl lived across town from me and as I began to accelerate across the freeway entrance, each light I passed would go out just as I was passing it. This was invariably on evenings when we had had sex. On other evenings some lights would go out but not like on the ones when our passions had been aroused. [38]

A young man from Harrogate, England, recalls:

> My experience of SLI started when I was 19. I was walking back from my girlfriend's house after a night out at the pub followed by a smooching

> session at her house. It was summertime, in the small hours of the morning, dark and quite cold. I was walking through open parkland lit by ornate old-fashioned streetlights when the nearest light post to the footpath went out as I approached. I thought nothing of it except the bulb has gone. The very same thing happened the following night. Again I dismissed it as a dodgy bulb and thought nothing of it. Well, it happened the next time I passed the light-post (similar conditions, i.e. early morning, dark etc.) at which point I started to think something was "up." Over the months as I returned home from my girlfriend's house the light-post would always do the opposite as to its original state, i.e. when I approached if it was off it would turn on and vice versa. After passing the light-post it would usually revert to its original state. [178]

This seems to be a clear example of the hypothetical model we proposed at the end of chapter three, whereby the same force field will put a light out if on, but re-light it when off.

But sexual unhappiness can trigger SLI as readily as sexual euphoria. Indeed, the event that first kindled my interest in SLI was at a conference at Hasselt, Belgium, when after my presentation a man came up to me to tell me of his experience: After difficulties with his girlfriend, the streetlights would go out outside their home. [209] A Welsh engineering student reports that he no longer does SLI, but that he did intensely over a three-year period when he was a teenager:

> The main time was when I was 16-17 when lights would go off most regularly. This was a time when a close relationship with my girlfriend was deteriorating and breaking up and when I

> decided to part with Christianity as I began to see it as hypocritical.
>
> At first after the novelty had worn off, lights going out started to irritate me a little, especially as I was usually in a bad mood at the time. Eventually, however, it got quite comforting and I started to experiment a little. I found lights would go out at will although whether I was simply pre-empting the event or making it happen I'm not sure. [92]

His experience is paralleled by that of a young lady from Leipzig, Germany, who reports:

> During the time when I took leave from my boyfriend, nearly daily streetlights went out, mostly just one, two and mostly when I went past on the opposite side of the street. After some weeks the phenomenon faded, but came again from time to time, when I did some flirting from men in the open. In the last years I triggered it when I have been excited either from worries, tension, or from fears. All in all around 90 times. [94]

It's interesting that only one informant mentions feeling terror. [20] This seems to be telling us that SLI is not a "fight-or-flight" response; when SLI happens, the SLIder does not feel driven or even encouraged to do anything. SLI is, as it were, self-contained, something that occurs but with no build-up and no consequences. This concurs with our finding in chapter three, that SLI is a notably trauma-free occurrence.

Here are two more very specific reports involving emotional states. The first is from Severn, Maryland. A physicist/engineer had a sudden surge of SLI when his mother visited, causing considerable domestic stress ("very high negative stress"). "One entire fluorescent light was swung round 90 degrees overhead, hanging by two chains – completely baffling at the time." [80] Similar incidents occurred in the Rosenheim poltergeist case we shall consider in chapter eight.

The second case is from Hastings, England. An art student reported that SLI happened to her after a painting class at her art school, which left her in an emotional state. [134]

As you see, this pretty well covers the entire range of feelings of which we human animals are capable; no wonder we haven't been able to link SLI with any particular frame of mind. Since these are all subjective states, it could well be that people are finding different words to describe what are essentially the same states. But these are the words SLIders actually use, and they couldn't possibly confuse "feeling happy" with "feeling unhappy," nor "elated" with "depressed."

So have we established that you need to be way up or way down to do SLI? No, by golly, we haven't! Just to keep us wondering, there are some SLIders who find no correlation with any particular emotional state. A correspondent from Oregon reports: "I have zapped them mad, in love, lost in job thoughts, day dreams." [2]

And a musician from Colorado finds:

> There is absolutely No tie-in with anything, not with moods, circumstances, job, weather, nothing. Completely random. Each time it happens I try desperately to figure out what common thread runs between happenings – as far as I can tell, there is no thread. [33]

We should also perhaps take note of the feelings SLIders express *after* their SLI experience. Most are merely puzzled. A London, England, psychologist was "amused and bemused" [37] but some have a more positive feeling such as this one from New York:

> I must admit, when a light comes on when I come close to it, I do feel excited. I can't explain why, it just feels as if I have more power or something when it happens. [31]

This is a rare observation by a SLIder who *feels* the force field we are hypothesizing. The great majority of SLIders don't seem to feel out of the ordinary in this respect, whereas we might well suppose that giving out that much energy would cause physical feelings of some sort.

In sickness or in health

In the previous chapter we considered the possibility that ill health could affect a SLIder's SLI-proneness. We need to distinguish, though, between *chronic* ill health, such as the SLIder who suffers from Multiple Sclerosis, and transient states such as migraine. SLIders rarely mention their state of health/ill health, but a few cases suggest that the *actual* state of the SLIder's health may induce a state of mind amenable to doing SLI. A Hungarian informant said whether he was in good health at the time but mentioned "a pain in his head," though he didn't elaborate or say that he also felt the same pain at other times. If this was the only occasion he felt this pain, this might tell us something, though no one else has mentioned anything of the sort.[4]

Some SLIders mention a possible connection with migraine:

> *Central City, Colorado:* If I am pre-migraine, for about 2 days before I randomly set off circuit-breakers, cause the phone machine to malfunction. [17]

> *A man from Augusta, Georgia, wonders:* I suffer from chronic headaches. I sometimes wonder if the lights are the culprit. [41]

Culprit or consequence? Could be either. A technician from Piano, Texas, dates the commencement of his SLI experiences to a very serious accident with an electric welding-machine, which paralyzed him for two weeks and left long-term damage. Since then he has experienced SLI more than 100 times, witnessed by family and friends though they dismissed it as coincidence. [10] A hundred coincidences and a possible cause?

An American cab driver feels he can pinpoint precisely – and quite dramatically – when his SLI began:

> The first SLI was preceded at around 8pm by an unexplained phenomenon I call the "Energy Ball," a source of strong electrical energy that appeared to follow the taxi I was driving. The weather was unstable, thunderstorms had been present. At a stop sign I observed a blue aura in the branches of some trees and the presence of an intense sparking sound similar to that made by a Van de Graaff generator. The phenomenon suddenly appeared to become attracted to my taxicab as if it was trying to overtake me. I accelerated and still it stayed with me for a moment. I could feel static electricity in the air. Then it slowly faded away as if I was leaving it behind.
>
> I have observed ball lightning before and there was no resemblance to what was experienced here. In addition, I had the impression that this phenomenon was well organized.
>
> I believe that my SLI activity was somehow related to the "energy ball" phenomenon. I hadn't noticed SLI occurring before the experience. [25]

The fact that this correspondent links BOLs (Balls Of Light) with SLI together is very interesting, for we shall be looking at this other anomalous phenomenon in chapter eight. Why there should be any association between the two anomalies is not easy to explain, but the parallels are intriguing. So, too, is the cabby's feeling that the phenomenon was "well organized." Yes, but just who or what is doing the organizing?

Stress resulting from pain is a further possible factor. An Englishman and his fiancée had previously noticed that he affected faulty lights:

> Last night I had just been to the dentist, and my teeth were in a considerable amount of pain. I stopped at a set of lights (which had accompanying streetlights) and both sides stopped working. At the same time, the orange signal light blew. We thought this was very weird. I mean, never before had we seen a good set of lights just... stop working. [213]

Crisis warnings

An electrical contractor from London, England, writes:

> The experiences have happened to me for about the last 15 years; it has been witnessed by my wife and countless friends. I first noticed this about 1980, when my best friend died, aged 20. And it has happened ever since. I have noticed when the streetlight goes out, it is like a warning that something is going to happen, an advance warning. My father had died after a light had gone out and ten months ago the last time I have experienced the streetlight going out my cousin aged 20 died. [142]

Now we are really edging into psychic territory, but if that is what this SLIder says he feels happening to him, we must take respectful note of what he says. This, of course, has important implications.

For while the phenomenon is still associated with the individual, it appears to be triggered by *something that is going to happen*, rather than the immediate action of the individual's presence. There is no reason for us to think that a premonition was involved, nor stress, for two of the three people mentioned died at an unnaturally early age, presumably unexpectedly. However, because no one else has mentioned this kind of a tie-in, we should perhaps play it safe and treat this case as coincidence pending further indications of the sort.

In the following chapter we shall note two other SLIders who claim to have a form of foreknowledge that a SLI-event is imminent. [15, 129] If we knew how and why this happens, it would guide us in understanding the interplay between the physical and the psychological forces involved.

Holiday mood

Although the majority of SLI occurs in everyday situations such as traveling to work, a good number take place when the SLIder is on holiday. Thus a Dutch informant tells us:

> I had SLI for the first time in Italy. During summer holidays in Rimini, only one streetlight in a row of many went out. One of the next days it happened again: same light in the same street. Recently it happened during a holiday week on the island of Rügen, Germany, only once." [162]

It is likely that people are more alert to what's happening around them when they are on holiday, but equally, it might be that the holiday mood creates a heightened degree of excitement. The writer does not mention doing SLI in his/her own country, but intriguingly adds: "I have the same experience and probably will have in the future."

A SLIder who put out not only streetlights in his native London, but also lights in shops and supermarkets ("only one aisle, fortunately"), also had an experience while holidaying in Funchal, Madeira, when "one of the fountains stopped suddenly as I walked past; none of the other fountains was affected." [172] Presumably the fountain was electrically operated.

On the other hand a SLIder who put out a row of lights night after night while holidaying in Tenerife never did it at home or elsewhere. So it seems likely that he was in an exceptionally exuberant holiday mood. [77]

Drugs and drinking

> *A pharmacology student at Bristol, England:* Andy and I regularly visited Clifton for a night out. I noticed that, during these forays, I became aware of occasional streetlight failures that happened to coincide with my walking beneath them. Andy also noticed this and we joked about this between venues. However, on one particular evening it truly spooked the pair of us. We had been walking up towards Victoria Square en route to the Albion pub when approx 3 streetlights thus failed. I was ribbed about this. We then walked through the path that crosses diagonally through Victoria Square. As I passed the 3 or 4 lights in this short section of pathway every light failed exactly as I passed beneath it. Both Andy and I were taken aback, and sobered somewhat. There was certainly no simultaneous failure, and they all flickered back on as I moved to the next light in sequence. [196]

The informant tells us that they had already had some drinks at that stage of the evening. No other factor seems to have been involved, and we can hardly suppose him to have been stressed as the pair of them migrated from one pub to another. So was it the drink he had already consumed? Quite a number of SLIders have mentioned drinking in connection with SLI, but only as one of several factors. A reasonable surmise would be that alcohol in some way facilitates SLI, possibly by relaxing the SLIder's mind, but it doesn't seem to *cause* SLI, just helps it to happen.

A SLIder from San Francisco writes:

> One thing which might be of interest to you is that all or most all people that I have mentioned this to are recovering alcoholics/addicts, and among this circle of people the amount of those with this ability is very high. [87]

There are some indications that alcoholic drink can indeed encourage or facilitate SLI. A SLIder from Birmingham, England, reports:

> The only time I have ever done it at will was when I was out with a friend. We had been to the Tyburn House pub where I had got fairly twisted. As we were on our way to a cinema, en route is a retail park with a collection of lights that are particularly prone to going out. For a laugh I started "shooting" them. To my amusement they went out as I "shot" them and came back on on request. This has never happened before or since. [177]

Alcohol is, of course, a relaxant, so it may be a way of getting round this curious reluctance of the lights to do as the SLIder wants them to. But then, it's the SLIder, not the light, that's doing the drinking. All the light needs is a drop of oil from time to time.

Few SLIders have mentioned drugs or any other intoxicants, but this may simply be natural reluctance. One woman mentioned that she experienced SLI after taking LSD when she was 19. This is a line of research that might be profitable – and would surely be entertaining – to investigate further.

How unconscious is SLI?

Apparently SLI happens out of the blue. If the individual is responsible, he does not know it but we have a few clues to the contrary. One SLIder tells us she has a premonition before she does SLI. And a few SLIders say they somehow *know* that they were responsible.

Assuming – though this is only an assumption – the subconscious mind of the SLIder is responsible, then from what we know of the mind we can speculate that somehow deep in its recesses is the knowledge that it has done SLI. To the best of my knowledge, no one has ever hypnotized a SLIder to see if in fact they have any subliminal awareness of doing it. It is certainly worth trying, even if the result is negative; that would tell us that the SLI-action is an automatic one, not willed or sought by the individual at however deep a level. Most things that people do are *willed*, consciously or unconsciously. So answering this question, establishing whether SLIders are in any way willing SLI to occur, would be instructive.

To all appearance, *the lights don't want people to think about them.* Put like that, it's absurd, but consider this comment from a SLIder from Denver, Colorado, a student at the time:

> As I walked across the campus, along a path lit by streetlights, several of them would go out as I passed under them. Oddly, once I began to change my thoughts to the lights going out, they would stop going out. [197]

So it seems that the streetlights refuse to perform not only when the SLIder *wills* them to come on, but even when the SLIder merely gives them his attention. What other appliance behaves in this uncooperative way? Imagine if your car or your dishwasher refused to start until you turn your mind to something else!

Surely there is a crucial clue here, if we can only interpret it correctly. We shall return to this tricky question in our final chapter.

Finally, what we are forced to do is acknowledge a difference in the SLIder's state of mind when she is doing SLI, and when she is *trying* to do SLI. Generally, she can't do it to order, which suggests that some kind of change has taken place. Has she stopped being angry, depressed, or elated – or whatever it was that triggered her original SLI? No, but perhaps the focus of her mind has shifted. Her attention is now fixed on the SLI process as such, rather than on whatever caused her emotional state. This is clearly demonstrated in the case cited at the start of this chapter, where the angry boyfriend stops stomping around campus and gives his attention to the SLI-happenings, whereupon he gradually stops doing it. If we could figure out what precisely has happened to his state of mind, which enables him to zap lights in one state, not in the other, we should be a major step towards understanding SLI.

Chapter Seven

From radios to railway crossings

London, England: Just a spare moment of your time to ask a quick question. I'd appreciate your feedback on a situation me and my wife have, whenever we argue, our television set turns on like its on standby mode, however, it comes up as a white screen for a minute or so, then turns itself off as soon as we stop for a minute and look at it. If the argument continues, it turns back on and then off again. This has happened on at least 5 occasions to my knowledge.

Recently we argued quite badly, and the TV came on downstairs, the TV in our bedroom came on (where we were at the time), the overhead light started flickering and as the argument progressively got worse, the light flickered more violently. The light bulb blew by the time we had finished and both the TVs had turned themselves off as soon as the bulb blew. (The blowing of the bulb caused us to stop arguing.)

There's also been occasions where we have been sleeping and one of us has had a bad dream and been woken up by what seems to be my alarm radio turning on, and also making the similar sound as if a mobile phone was placed right next to the alarm.

I'm now at work and done a few searches for sites regarding the supposed phenomena ''SLI" or "sliders." Do you think that this may possibly

> be affecting me or my wife? Looking forward to your reply. [194]

I might have replied that perhaps it would discourage him and his wife from arguing so much; instead, more politely, I said that this sounded like SLI in another form. The previous chapter leaves us in no doubt that the SLI effect is at least partially psychologically conditioned. SLI happens when the mind of the SLIder is in a certain not-my-usual-self condition. When we discover that this is no less true when other appliances, instruments, or machines are affected, we can be in little doubt that the same force, or one closely related, is being exerted. A much-troubled woman from Oklahoma reports:

> In a three-week period one time I had an iron, coffee maker, tea maker, oven, hot water heater, watch, clock, telephone, typewriter, light fixtures, outside lighting, computer and car problems happen. The electricity had no strange power surges in our home – and it's *expensive!* [104]

She was particularly unlucky, but quite a few SLIders report that they also affect other types of appliance. Those most frequently reported are electrical or battery-driven appliances, but it seems that sometimes other machinery, such as compasses and clockwork devices, may be victimized. We shall see in the following chapter that some so-called "electrical" powers can affect materials that are commonly thought of as utterly resistant, such as wooden furniture. So just because most of the objects that seem to respond to SLI are electrical, we should not conclude that the force itself is electrical. To take a simple analogy, the force required to turn on a domestic light switch is manual, even though the effect is to activate an electrical process.

So though we are primarily concerned with people who affect streetlights, reports from people who affect other appliances, such as the following instance, even if they do not do SLI as such, may contain useful insights.

Electrical appliances affected by SLIders

> *Student from Stourbridge, England:* One summer evening when I was 17, I took the family dog for a walk. It was still daylight. As I was walking back home up a very steep hill, a car parked by the side of the road (on my side of the road) flashed its headlights at me. As I approached I peered into the car in case it was anyone I knew. The car was empty. When I got home I told my father (at this time an expert on cars) and together we drove past the car to see if (1) it would do it again (it didn't) and (2) was it a "trick of the light." I put this experience to the back of my mind.
>
> A week or two later I was walking in a nearby street (on my way to college via a canal short cut) when another car parked by the side of the road flashed its headlights at me! Once again it was empty.
>
> The final experience occurred about a year later. I was in my final year at college and had stayed late with friends to finish off some work. It was winter as it was dark when we left. As we walked through the car park at the rear of the building, I jumped out of the way as a nearby car's white reversing light came on. I turned to look into the car (intending to give the driver a "dirty look" for not checking before reversing) but once again the car was empty. The incident did not appear to have been noticed by anyone else. Since then I can honestly say I have never had a similar experience. [195]

No, it's not SLI, strictly speaking. But it is surely something very like SLI. No other SLIder has reported blinking headlights. What did these three empty cars have to say to our SLIder, and why did they say it in so obscure and enigmatic a fashion? What force can make this happen? Who or what decided to do it? If this was indeed the SLI force doing something else for a change, why did the student affect the cars, but leave the streetlights unaffected? The questions come tumbling, but we have no answers. It's hard to believe people can possess two different kinds of force, one that preys on streetlights but another whose victims are car headlights. Yet it seems that's just what happened.

And it's just as puzzling on the far side of the globe. What conceivable rationale underlies this classic instance of machine busting from an Australian bank employee?

> The experiences I have been involved with are different from SLI but still relate to electricity. I work in a bank with a great deal of electronic equipment. In the first case, I was working as a teller on the counter of a very large, busy branch. The terminal would regularly go down for no apparent reason. Technicians were called in. After a few months, swapping machines, people, positions, it still occurred. But only to me. The position where I worked or the machine I worked on seemed to make no difference. Others were put on my machine and I on others. No one else had problems. Whichever machine or position I used, the computers regularly went down. In the end the technicians came to the conclusion that there was nothing they could do and it must be something caused by the operator. They admitted that they had never come across anything like this before. We learnt to live with this at the branch. I was scheduled to do other jobs whenever

the machine went down. It was frustrating for everyone, though, because we were a very busy branch. I was especially embarrassed.

Shortly after, I asked for a transfer. The new bank was tiny, only five staff. My machine was on the end [of the counter]. This means the security system, the printers, and main cables were in this region. Over a two-year period, the PIN-pad device (electronic key pad for customers to type in their secret numbers) kept going down. Obviously this was an inconvenience. The computer had to be downloaded and checked. Eventually, a steel plate was installed to prevent interference from the equipment. (Other branches had reported similar problems but they were soon solved with moving cables out of each other's way. That did not work here.) After fitting the steel plate, we still had problems. I use all three terminals, but only have problems with the end one. Occasionally others use the end terminal, but no one else had any difficulties. The problem is on-going. The technicians have now replaced all the cables. No mention has been made of my interference nor have I mentioned it. They are different technicians who come to this branch. Until I read about SLI I did not think anything of the experiences.

A third experience occurred at home and it has crossed my mind as to what effect I have and why. We have a system called a "video sender" whereby a video placed in a recorder can be viewed on a television in another room. The system works well until I stand in its path. Then, the interference ruins the reception. With anyone else, even groups of people, nothing happens.

> We have had parties with a room full of people where the video and video sender are, and another room farther away where the reception was being picked up on another television. There is no interference until I cross the path. This is very frustrating. [141]

The fact that this woman affects quite different appliances both at home and at her workplace, and that in neither of these locations do other people have the same effect, tells us that it is unquestionably an individual force, hers and hers alone, unrelated either to the environment or to other people. Her difficulties have evidently been going on over a long period, which seems to rule out short-term emotional states such as depression or stress – neither of which, in any case, she mentions as a possible cause. Her problems seem to relate to a *chronic* or at any rate prolonged psychological condition, whereas the Stourbridge case cited earlier relates to a short-term state that never recurred.

Indoor lights

Our household lights misbehave all the time, it seems to us, so it is only when their misbehavior becomes striking that we let it bother us. It happens. For instance:

> *Woman from Hickory, North Carolina:* I notice when I get aggravated, fluorescent lights go out over my head, or over the heads of those with whom I am upset. [1]

This is quite unusual. Since most of us spend much of our time under fluorescent lights, particularly at work, such as hospitals and offices, chaos would result if SLIders continually affected the lights.

We have received several replies like this one from a housewife in Birmingham, England:

> Although I have never caused SLI (to my knowledge), I seem to have the same effect on the lights in our house – quite frequently, when I turn on a light, it blows. My husband complains at how often he has to change the light bulbs. This does not happen all the time, only when I'm "charged" with emotion. In fact, sometimes I can feel when it is going to happen! [129]

A woman from Minneapolis says she affects "…very frequently lights in my home or friends' homes." [6]

In the preceding chapter we cited a woman from Leeds, England, who when her partner left her, had to replace virtually every light bulb in the house as she successively blew them one by one [70], but there isn't always so obvious an emotional cause. In the following instance, the SLIder makes no mention of being in any special kind of state:

> *Gainesville, GA:* Once or twice a year, year after year, light bulbs – some virtually brand-new – have come on briefly and then burned out as soon as I switched on the controlling wall-switch. In virtually all these instances, such anomalies as power surges or malfunctions could easily be eliminated. Moreover, in the majority of instances I knew unmistakably, an instant *before* I touched the switch, that the bulb would burn out; before my fingers touched the switch I knew, clearly, what was about to happen. At no time was my anticipation wrong. [15]

The "oh, here we go again" feeling seems to happen more with domestic lights than with public lights, but that may be simply due to circumstances.

This next case is especially interesting:

> *Woodland Hills, California:* I first noticed this phenomenon when I moved to New York City. I lived there for five years in 3 different apartments, and my roommates always referred to me as "the fuse blower." When I would come home from work, usually irritated by a crowded subway ride, the first light I would put on would "pop" and be burned out. Also other appliances wouldn't turn on when I touched them, and someone else would have to turn them on.
>
> I worked in an office then, and one day the woman in the office next to mine called me over and really made me angry. Suddenly, all the lights in her office went off, and the switch plate was too hot to touch. The maintenance man was baffled because she was on the same circuit as the rest of the offices, and only hers went out. I told her that I did it and she never bothered me again. [9]

This seems to involve an electrical "impossibility." Those of us who use the stuff learn that nothing about electricity is impossible, but in this case even the professional was baffled! The fact that the office episode involved interpersonal relations is also unusual; SLI is generally a private affair between person and light, and it is very rare for other people to be involved as directly as this. Office workers take note: before bawling out a colleague, it's prudent to make sure she's not a SLIder!

Several other SLIders claim to do domestic lights occasionally, but by contrast there are others who specifically assert that they cannot affect any lights except streetlights. [13, 22] As if the SLI enigma wasn't puzzling enough on its own, it is hard to imagine any possible basis for this selectivity, which sometimes uses the force to affect many kinds of light, while for others it chooses only one kind. You would think that, compared with knocking out a streetlight, terminating the career of a domestic light would be child's play; perhaps the SLI-force doesn't always find ordinary household lights enough of a challenge?

Shop or store signs

In the previous chapter we noted a case where the SLIder extinguished a printer's sign that seemed to have a particular significance for his family. [140] The following informant does not say if he has any special feelings about McDonalds, for or against:

> *Portland, NY:* Once I was pulling into a McDonalds with a friend who had been with me on several occasions when I put out streetlights. The McDonalds sign out front was unlit although it had been dark for several hours. But as we passed by it suddenly flickered to life. [78]

The same informant also wrote that once he was in a parking lot waiting for his fiancée "and getting fairly angry," "when suddenly every light in the parking lot went out (and a few store lights) at exactly the same time."

But SLIders who do shop signs are rare. More typically, an Australian couple report:

> We never observed this effect with traffic lights, illuminated signs or anything but streetlights. [126]

However, they report another feature that seems to be unique. Intriguingly, they tell us that SLI occurs only when they are driving together, never when either of them is driving alone. We do not know of any other instance of this, which seems to imply that they need the *combined* force of both of them together for SLI to occur.

Checkouts and cash registers

Standing in line at the supermarket checkout is tiresome enough with the lady in front of you fumbling for her credit card, without having to put up with her SLI-force as well. Both these SLIders have found practical solutions to the difficulty:

> *A counselor from Fort Myers, Florida:* About 50% of the time I approach a supermarket cash register it malfunctions, and the usual comment by the operator is "It never did that before." My children took this as a matter of course and would get angry because they could not get out of the store until the problem was over. What we do now is that my wife takes over and I move about 15 meters from the machine – this seems to be about my "range." Then the machine resumes functioning. [47]

> *A Canadian housewife who also does streetlights:* The other thing I seem to affect are computer tills. I will be in line and the till will malfunction; if I'm with my daughter she will pull me back a bit and the till will function normally again, always with the comment from the cashier, "I don't know what went wrong: it's perfectly fine now," or words to that effect. This doesn't happen all the time: I've noticed it is if I'm worrying about something or distracted. Then when I concentrate the till will function again. [205]

Note that this is contrary to the general experience with lights. Yes, they may go out if the SLIder is distracted, but they don't like it when the SLIder concentrates her mind and, as a rule, will refuse to play ball.

Should we think of these incidents as connected with doing

SLI? We can't afford to miss any clue, however fringey. An American SLIder relates this incident:

> There is a mischievous side to it sometimes, like someone or something or some part of me wants to add to my fun if I am high. Once in a bubbly mood I went to a liquor store for a bottle of scotch since we were camping and a drink would be appropriate. I paid for it with a $20 and the cash register told the clerk to give me $83 change, which he started to do, but then laughed and said the usual "It never did that before." [47]

Computers

Another favored target are computers, and they share with domestic lights the strange feature of foreknowledge:

> *Eugene, Oregon:* There have been several times when I knew my computer was going to die, and need repairs – just before it died and showed no outward symptoms: did I sense it or cause it? [2]

Whether those who have this happen to them are really seeing into the future, or whether it is simply a piece of odd behavior in the brain akin to *déja vu*, is an intriguing minor puzzle in the SLI enigma. If SLIders were responsible for their actions, then we could say that they were subconsciously aware of their intentions – that one part of their mind is aware of what another part is just about to do. Until we know whether, and to what extent, the mind is responsible for doing SLI, we can't say whether this is genuine foreknowledge or even a warning premonition.

Several others say they affect computers and word-processors. [3, 6, 50, 56, 61, 72, 166] Unlike streetlights, computers are often felt by their owners to be essentially *personal.* While many of us curse our computers when they misbehave, others feel that there is some kind of love-hate relationship going on, a kind of affinity, and I am surely not the only computer-user to feel possessive about the contraption on which I am writing this book, and resentful when other people monkey with it, even change its settings. In particular, as with cars, there are people who are "good with computers." Our "computer-guru" at my workplace is such a person, giving it a lesson in good behavior when it acts up. An engineer from Brighton, Minnesota, is fully aware that he possesses this valuable gift:

> I have the experience of turning off streetlights frequently; I also have a kind of "healing" effect on computers. In our research department I am frequently involved in helping colleagues with high-performance computer installations. When they have trouble and ask me to come by and help, the problem almost always disappears when I handle the machine – I've acquired a somewhat humorous reputation over it! [63]

It is reasonable to think that his ability to turn off streetlights and this effect on office computers are related somehow. If so, this must be one of the rare examples of SLI being put to good, practical use.

Games

With computer and other electronic games so popular these days, it is perhaps rather surprising that SLIders rarely mention them. But we have had a few letters such as this from a college graduate from Carbondale, Illinois; after telling us of her SLI experiences she adds:

> This might sound really crazy but there's something else I've noticed. When I play pinball, which is not too often, I have a lot of trouble with the ball going straight down the middle of

> the flippers towards me. It happens nearly every time; therefore I don't play pinball that often. I know that pinball machines use magnets in some way, but I'm not sure about the scientific aspect of it. Could this be somehow connected with the lights going out? [161]

A SLIder from London, England, mentioned gaming machines [48], but he also claims to affect *dice*, where no electricity is involved. Joseph Rhine[7] and others have tested psi-ability by asking their subject to affect the throw of dice with their mind, but there has been no previous suggestion that it is linked to SLI. This opens up an intriguing line of investigation, perhaps involving a visit by SLIders to Las Vegas. If any such connection should be established, it might invalidate years of dedicated work by psychical researchers. On the other hand, it might open the way to new dimensions of the mind in which a seemingly mindless activity is replaced by one in which choice, direction, and intention play a part. (It might also suggest ways in which the gaming establishments should redesign their operation!)

Elevators

Many people have a phobia about elevators – my wife and I once walked down several flights of stairs in a Paris building with director François Truffaut because all three of us shared it – so it is perhaps comforting to know that few SLIders find their ability includes elevators or lifts in their repertoire. But sometimes they do. A father from Fort Myers, Florida, after describing his own SLI experiences, recalls this unnerving episode:

> My son who is 20 does the same; when we are together the interference rises to about 75 percent of the time. Once I had to take him to a hotel in Miami for a high school conference. We entered a posh hotel and found we had to register

> on another floor so we went to the elevator, one of 8. We entered, the doors closed, then opened again. We waited but nothing happened. When we stepped out we noticed all the elevators were back on the first floor [i.e. ground floor in English usage] and were opening their doors. A voice came over the PA saying a sensor was recording a fire somewhere in the building so the elevators would not function. A few minutes later the PA announced there was no fire but a sensor in the first floor elevator wing was malfunctioning. Seeing the possibility that we might be doing this we left the wing and in a moment the elevators were operating again. We took the stairs. May be coincidence, but it happens far too many times to me. I walk into the bank and the computer goes down at that moment. I walk into McDonalds and their computer system goes down. I have learnt to stay back from the counter a meter or so. [47]

It is logical to suppose that it might require the combined SLI-power of father and son together to affect a bank of eight lifts, except that the logic of SLI isn't always so evident.

Traffic lights

Using SLI to break the bank at Monte Carlo would be both scientifically gratifying and financially satisfying, but perhaps the most useful everyday manifestation of SLI would be the ability to control traffic lights. A few SLIders have reported that this happens for them, though none has been able to do it on a regular basis. A Canadian SLIder says he has only once done a traffic light [83] but a New Yorker has done many. [143] Once, after giving a talk about SLI in north London, I drove home – some 15 kilometers – and as I negotiated the streets, every single traffic light was either green, or turned green, as I approached. Coincidence, maybe, though I like to think that whoever controls SLI did it for fun or as a thank-you to me for giving a public talk about it.

A London housewife recalls:

> At one time I was able on a regular basis to make traffic lights change sequence simply by expecting them to do so. As my children were small, I used to "zap" [the traffic lights] with my hand/mind on approach and they would ... [turn green], out of turn. I do not have this ability any more, sad to say, nor can offer any explanation of how or why it was possible. However, my children can clearly recall the occasions when this happened.
>
> Is this any different from something I still do, which is to know when another vehicle is approaching (in daylight) on say a narrow road which has bends round which you cannot see? Our lives/accidents have been saved/prevented by my stopping the car on numerous occasions only to encounter seconds later a vehicle appear with no other apparent warning. [143]

Note that this is another exception to the "rule" that a light will refuse to do what the SLIder wants it to do.

Railway crossings

The good people of Fredericksburg, Texas, are warned to steer clear of one of their fellow citizens:

> I have affected railroad crossings, when the barricades would come down, and the lights and bells would come on, with no train, workmen or other people around. [60]

This would appear to be a fortunately rare example of SLI being potentially dangerous. Train crossings are also affected by a woman from Woodville, Texas, who adds, "It seems to happen only when I am very upset."[76]

The Fredericksburg residents may be reassured by the fact that the SLI – if SLI it is – takes the form of signaling a train when there is no train; vice versa would be truly alarming, with the gates opening just as the westbound Twentieth Century is signaled.

Other types of electric appliance

Sooner or later mysterious forces seem to intervene in the operation of every kind of gadget, from the domestic to the public:

Aircraft electronics:

> *Central City, Colorado:* When I am in electronically sophisticated cars (electronic ignition, door locks, window openers, etc.), I often cause that to malfunction. Car won't start, or doors lock and won't open. I keep a simple house, needless to say. I have an old Apple computer that is "used" to me. My cars are simple, though at first I have to coax them along. After they are used to me they are in some way extensions of me. I had one truck 13 years and it would start under any conditions for me. This truck started without a key for me and no one else could do this.
>
> When I fly on airlines, I become aware that the pilot often said there would be a delay to check out an unknown electrical disturbance. If I meditated and shielded myself, in a few minutes the pilot would announce there was no apparent trouble. So I now routinely shield myself with color imagery so I can pass through the safety check points, and then when I am first on the

> plane I quiet myself and shield again. [17]

The mention of deliberate calming raises this question: To what extent can SLI itself be blocked or otherwise manipulated? From what we have seen, hardly at all, but evidently some SLIders have found ways of handling the awkward situations that SLI can sometimes land them in. Potentially, SLI could make life extremely tiresome, not just for the perpetrators but for the public at large, by causing a shutdown at an inconvenient moment. But fortunately few SLIders have mentioned anything of the kind; and if it happens frequently, perhaps they too find a way to manage it.

Alarms [29]

Business pagers [27]

Cameras:

> *Falls Church, Virginia:* I caused cameras to stop working temporarily when I strenuously objected to my picture being taken. [46]

Car electronics [6, 17]; *Car alternators*: "I am now on my eighth." [73]

Clocks [21, 48, 56] *and watches* [27, 34, 46, 48, 54, 56, 86, 166]:

> *Pardeeville, Wisconsin*: I can't wear a watch – the batteries only last a quarter of the time they're supposed to. [54]

> *Brooklyn NY*: I also kill wacths. [sic] [20]

It would be interesting to know if this includes hand-wound watches as well as battery-powered. When my mother died (peacefully in her sleep at 95), her bedside clock died at about the same time, but so far as I know she never did SLI in her life. Many instances of this synchronicity have been reported, but they are usually dismissed as folklore, so it was disconcerting to have it happen within the family.

Copy machines [3, 166]

Floodlights [62]

Hair dryers [85] *and Irons* [37]:

> My first knowledge of myself as an "electric person" was at age 21. I was ironing a silk dress and had the iron on *low*. My mother was arguing and baiting me and I was holding my tongue. The face on the iron literally melted and silver liquid metal spilled down on my dress. I know for certain repressed anger contributes to my ability to disrupt electrical equipment. [17]

Lights in refrigerators and other domestic appliances [86]

Lottery machines [42] *and Video games* [43]

Microphone systems [61]

Radios [19, 52]:

> *Alexandria, Virginia:* I cannot sit near or tune a radio when I am agitated. If a radio is in tune when I am next to it, it will be badly out of tune when I step away. And vice versa. This effect is so pronounced and occurs so frequently that one friend dubbed me "radio-active." A similar effect

occurs around televisions. [19]

> *Bricktown, New Jersey:* Radios and TVs sometimes change volume or fine tuning when I am close to them. [52]

Store safes [3]

Telephones [17]

TV and Videos [11, 17, 52, 61];

> *Athens, Greece:* When I was a teenager my father would exclaim "Don't touch the TV – tell me what channel you want, I will change it." This was in the days before remote controls, and every time I got up to touch the TV, the picture became distorted with electrical "interference." It didn't happen with anyone else in the house. [61]

Typewriters [61] *and printers* [166]

You might think that SLIders go about the world bringing disruption and confusion wherever they go. Fortunately, these incidents are rare and would in the normal course of things be attributed to malfunction or misadventure, except that the SLIder's experience – and sometimes her gut feeling – suggests that there is more to them than that. But the ability to affect other appliances is not universal by any means. Some who do streetlights go out of their way to emphasize that they do NOT affect other things:

> *Taylor, Montana:* The weird thing is, I have never had any problem with any other electrical devices (watches, TV, radios, etc.) or any other types of lights. [5]

A SLIder from Fair Oaks, California, [7] reports the same. How can it be; how can the force that for some people affects TV sets and aircraft electronics, for others is content to play with streetlights? Yet again, the question poses itself, what's so special about streetlights? If a choice is being made, who or what is making that choice?

Non-electrical devices

Once, after a presentation on SLI, a woman from the audience came up to me and asked if I had ever had a case of someone who affects *bread.* When she and her mother went to the supermarket, her mother would approach a stack of loaves and it would start to shake, and if she didn't move away quickly, it would end by toppling over altogether. Once, they were standing outside Selfridge's department store in London (England)'s Oxford Street, admiring a splendid window display of different kinds of bread. Suddenly they realized it was wobbling; a moment later, the entire display tumbled and fell to the floor of the window. A passer-by who was standing beside them admiring the display took one look at her mother, and hurried away

The great majority of items on SLIders' must-do list are electrical, and we may reasonably suppose, first, that the watches and typewriters they mention are electric, and second, that it is the electrical parts of cash registers and the like which are affected. But dice and compasses have been mentioned. An English informant says that his brother, who lives in Florida,

> ...has described to me how from an early age wrist watches would not work on him. He states that sometimes mechanical watches will go haywire with the hands moving around the face at great speed while bedside digital clocks will lose periods of time during the night. [179]

For sheer mind-boggling bizarreness, it would be hard to beat this account from a Canadian housewife – the one cited earlier who affects supermarket check-outs:

> When I was at University and really stressed out during exams, whenever I passed my collection of music boxes they would start to play. I remember one in particular was a piano, which played on and off all night. I eventually threw it out into the snow, so I could sleep. [205]

She stresses that these were the kind that are wound by hand, not electric, nor powered in any way. She always kept them unwound to protect the springs, winding them up only to play them.

Only one SLIder, the lady from Athens, Greece, who knocked out the restaurant's electrical system, mentions compasses. [61] Similarly, dowsing rods, which we might expect to be easily affected, are cited by only one, from Norwich, England. [48]

Healing effects

Examples of a SLI-type force being used by one person on another are rare, as we have already noted. However, the same SLIder from Athens who does compasses also relates this experience:

> At dinner in a taverna, one of my clients said, "You do healing, don't you? Could you get rid of my husband's headache?" Normally I would do hands-on healing, but since we were in a public restaurant, I SENT the Reiki healing energy across the table, aimed at her husband. About 20 minutes later I asked him how was his headache? He said, "It's gone." Unfortunately so had the microphone system. It had been working fine beforehand, but when the cabaret tried to start up again, they just couldn't get the microphone

> system to work. It was out for the rest of the evening. [61]

Person-on-person healing has been extensively studied, but with conflicting and confusing results. The healers themselves add to the confusion with their claims. Some say their power comes from God, others from faith, others from psychic gifts of one kind or another. Yet others simply say that the power is in themselves – it's just a gift they have. Yet they all produce results – at least on some people, some of the time.

Everyday accidents

Many of the events described in this chapter might well be described as accidents, were it not for the circumstances – the fact that they occurred synchronously with other happenings, or happened persistently, or simply that the individual to whom they happened was convinced that she was somehow the cause of them.

This raises a question: How many of the everyday incidents that we suppose to be accidental could be, in fact, the result of SLI? We shall consider in chapter nine some suggestions that have been made that there are no accidents, as Freud famously proposed in his thought-provoking *Psychopathology of Everyday Life.*[8] Freud, of course, attributed the untoward happenings to psychological factors that could be brought into the open by psychoanalysis. But SLI-type forces offer an alternative, and in some respects simpler, explanation.

It is tempting to speculate whether, in fact, a great number of "psychic" happenings should be attributed to SLI-like forces. What about oddities such as this:

> *Hickory, North Carolina:* I once saw a rock glowing, and out of it came pinpoints of light – five or six of them. They rose into the air a few centimeters, then went back into the rock, after sort of inverting themselves. They did this three

times within, say, thirty seconds or so. [1]

When we considered the people who do SLI, we discounted "psychic" people. The great majority of SLIders are so very obviously *not* extraordinary in any other way than their SLI. But every one of us has weird incidents in our lives, like the woman from Hickory, for which no explanation suggests itself. It would be ridiculous to propose that whenever we can't find a conventional explanation, we should fall back on SLI. Unless and until we have reason to think otherwise, we should proceed on the assumption that SLI is a naturally occurring process capable of a scientific explanation.

Moreover, the fact that people who extinguish streetlights also do these other strange things – and surely the powers are related, if not one and the same – encourages us to look more widely at extraordinary experiences of the kind described in the next chapter.

Chapter Eight

Electric girls and glowing nuns

Angélique Cottin was no more electric than you or I are. It's what she did with her body electricity that earned her a label as "the electric girl" and her place in the history of science. What she did mostly was repel – or sometimes attract – items of furniture, including massive, heavy rustic chairs and tables, as well as a variety of household articles. Angélique's force was so great that three grown men could sit on a table, and yet it would be hurled away from her with irresistible violence. When, in the 1840s, this 14-year-old peasant girl was brought from her Normandy village to Paris to be examined by the eminent scientist, François Arago, he described what she did as displaying "a new force." That's pretty well what people are saying today about SLIders.

What has Angélique's strange power to do with SLI? Not much, other than this – both involve interaction between a human individual on the one hand, and on the other objects that are supposed to be inanimate, incapable of any behavior beyond what they are empowered to do by nature, in the case of Angélique's furniture, or their designer, in the case of streetlights. In this chapter we shall be looking at a selection of some anomalies that feature this interaction, which is, clearly, the distinguishing feature of SLI.

Scientists would like to find a box where SLI could be tucked away comfortably, preferably with a label derived from the Ancient Greek, where it would behave according to the rules of science, preferably by not happening at all. But denial has got them nowhere. Their conscience won't approve their simply looking the other way, or finding some other box that more or less matches SLI, hoping we won't notice. If SLI has rules – and we shall see that it has, sort of – they aren't science's rules. SLI is a nuisance, but then anomalies *are* a nuisance. It's how they make their living. (And, of course, explaining anomalies is how the scientists make *theirs.*)

So are we being hard on scientists, who surely have enough to do as it is with climate change and what may be happening on the far side of the universe? Compared with such issues, turning out streetlights seems small fry indeed. Our reply would be that they might learn more about the way the cosmos is managed if they would look harder at SLI and other such anomalies. (If all goes well, you will understand why when we reach the end of our last chapter.) For there are many other curiosities like SLI kicking around – bothersome anomalies, so far as the scientists can see, but potentially pregnant with enlightenment. Charles Fort, forerunner of today's anomaly-chasers, in his four marvelous books called them "the damned."[9]

In a world where hot stones come hurtling out of nowhere, where furniture wanders around empty rooms, where mysterious fires break out and mysterious lights flash, Fort would surely have numbered SLI among the damned. Many investigators, notably William Corliss, Vincent Gaddis, and D. Scott Rogo,[10] have devoted their careers to studying these events in which inanimate things interact uncannily with us human beings and interfere in our activities – as we, it seems, with theirs. Does the same strange force, which we have labeled the SLI-force, make these other puzzling things happen too?

Mostly, they are things we'd rather not have happen. Some are merely a nuisance, tiresome, rarely causing serious harm or damage, but interfering with our lives in ways we could manage without. We want our cars to start, our TVs to stay tuned, our hair dryers to dry our hair, just as we want our streets to stay lit. But not all of them are hostile, destructive, or troublesome. Sometimes, as we shall see, they can be beneficial, even life-saving. Which just makes them all the more puzzling. Many of them have been assigned paranormal explanations for want of any better. And indeed, when an endangered climber is guided to safety by a ball of light which morphs into a Buddhist priest and then, mission accomplished, shrinks back into a vanishing BOL again, it may well seem that nothing short of a paranormal explanation will do.

But no, we don't have to give up on science so easily. We ended the previous chapter with a strange experience by a SLIder who, understandably, supposed that it was somehow related to his SLI. Which it may well have been. But there are many people who have never spooked a streetlight in all their born nights, but who have experiences so similar to those that SLIders have, that we are justified in looking for common factors to account for the uncommon phenomena.

In the 1970s I investigated the experiences of a woman in the west of England who habitually saw what she declared to be UFOs. They were also seen by her daughters and her very supportive but skeptical husband – but only when she herself was present. She also caused other phenomena of a kind that have become familiar to us, such as making the TV switch channels of its own accord. I didn't know about SLI then, but if I was investigating her case today, I would take her out onto the street to see what she made of the streetlights, or they of her. Surely she, like many of the people described in this chapter, possessed something of the same strange ability as our SLIders?[11]

Angélique's extraordinary power first manifested at 8pm on 15 January 1846. With three other girls she was sitting in Madame Loisnard's house in the village of Bouvigny. They were occupied with making silk gloves, a local cottage industry. Suddenly the oak frame, which held their work, was overturned, and their light thrown across the room. Arguing among themselves as to who had done it, the girls put things right and started working again. Then it happened a second time.

This time they paid more attention, and it became obvious who was responsible: Madame Loisnard's 14-year-old niece, Angélique. Nothing happened unless she was there. But equally clearly, she wasn't doing it on purpose. Was she bewitched, they wondered?

The village curé pooh-poohed the idea of witchcraft, but offered no alternative explanation. Yet if it wasn't witchcraft, what was it that could, for instance, hurl a heavy chair across the room even though it was being held firmly down? When Angélique was in the "electrical" state, almost anything that touched her apron or her dress would fly off, however tightly held. Maybe, the curé suggested, Angélique was suffering from a rare, perhaps unknown illness. He advised calling in the doctor.

The learned men of the community observed and recorded the phenomena, but could not account for them. In due course Angélique was taken to Paris to be studied by a commission headed by Arago, probably the most distinguished French scientist of his day. They were puzzled by the paradoxical nature of the phenomena; in so many respects they seemed to be electrical, in so many others – notably the effect on wooden furniture – they evidently weren't.

Yet Arago noted that when he touched Angélique during her paroxysms, he felt a shock as if he had touched an electric battery. Moreover, if a magnet were placed near her – even without her knowledge – she would at once start to tremble violently. Such effects point to electricity, but on the other hand she did not affect a compass needle in the slightest. Moreover, most of the objects she affected were wooden.

Though she was awarded the "electric girl" label, Angélique was not the only person to display such effects, though she was the most thoroughly studied. A few years earlier, in November 1839, two Turkish girls came to France hoping to make a fortune by displaying their ability to make huge, heavy wooden tables move, though the girls stood one to two feet away from them. Though their remarkable ability had first occurred spontaneously, once discovered, the girls saw no reason why they shouldn't profit by demonstrating it – as indeed they did, for a while. By this time, of course, as with Angélique when she reached Paris, these demonstrations were the result of positive acts of will, not spontaneous happenings. For a few weeks the "Smyrna Girls" were the talk of Marseille, where they submitted to inspection by the doctors. Then suddenly their power deserted them and they could do nothing.

What had happened? Quite simply, the weather had changed, and with it the humidity. Their strange ability, which had flourished in the dry heat of summer, vanished as the weather grew wetter and colder. Angélique, too, may have been affected by atmospheric conditions, though in her case it seemed to be the humid weather of winter, not the dry heat of summer, which favored her powers. For a week preceding her first incident, the weather had been heavy and stormy, with many thunderstorms; the air was full of electricity.

It seems logical to suppose that the state of the weather was somehow connected with these strange powers. At the same time it is likely that their personalities had something to do with the matter. By all accounts Angélique was a dull, apathetic creature with little to say for herself. It may be significant that her "attacks" usually came on between six and eight in the evening. Perhaps the first outbreak was an unconscious protest at having to sit working in the evening? But though the state of the weather and Angélique's personality surely played their part, the nature of her power remained – and remains – unknown.[12]

SLI and Electric Girls

There have been dozens of so-called "electric" people, all of them displaying much the same sort of puzzling behavior as the poltergeist subjects we shall consider later. As far as I know, none of them exhibited SLI. You might think that 19th century Angélique repelling her chairs and tables is a world away from a 21st century SLIder zapping lights on the streets of Auckland, New Zealand.

But consider this: we have a lady down in Somerset, England, who not only does SLI but repels household items the way Angélique did, as well as displaying many of the "side-effects" reported by SLIders in the previous chapter:

> I affect at intervals and to varying degrees electrical equipment and appliances, metals, plastics and man-made fibers I either "attract" or "repel" exaggeratedly. Clockwork will accelerate or stop completely, broken or "dead." I blow fuses, make switches/meters/timers jump or re-set. I make magnets re-align. [206]

She relates her ability to do these things to ill health in childhood, and the treatment she was given, which "triggered my becoming 'charged.'" She has no doubt that her SLI-ability is just one feature of her overall condition, and we must consider the possibility that their strange powers are, if not one and the same, ay least closely related. Since our Somerset SLIder presents just the same power as Angélique and the Smyrna Girls, it's a fair bet that they, too, would have exhibited SLI if there'd been electric streetlights around in their lifetime.

Psychokinesis (PK)

The scientists have given people who do psychokinesis a box of their own and a label derived from the Ancient Greek. Big deal. Neither they nor the scientists who study them know what force is empowering them, or why the results are so small and take so long to bring about.

The experiences of individuals like Angélique, the Smyrna Girls, and others like them are naturally classed with those people – usually though not invariably adolescent females – who claim to be able to make objects move simply by focusing their minds on them and willing them to do so. The differences between the girls' uncontrolled ability and the PK subjects' supervised feats are obvious, since the latter are deliberate not spontaneous, and demonstrated in the laboratory not in an everyday setting. But the force is surely much the same.

The ability to make an object move, apparently by directing will power at it, is a phenomenon that has been extensively investigated during the past century.[13] There is no doubt that a few gifted people can do it, though not easily; waiting for a subject to move a matchbox across a table is not for the impatient. PK is generally observed under laboratory conditions; that is to say, it is done deliberately and intentionally, closely watched and filmed – in conditions very different from those in which SLIders function. To what extent PK occurs spontaneously is difficult to assess, though it surely happens because that's what brought the subjects to the laboratory in the first place. The objects that are moved are generally small and trivial, such as a key or a coin, and even with objects so small a considerable amount of effort is required. So spontaneous PK could easily occur unnoticed if no one is watching.

A curious variant of PK is thoughtography, in which the subject claims to project mental images onto film. Much of the evidence for this phenomenon comes from Japan, but the most dramatic thoughtographer was the American Ted Serios, who produced some striking yet not entirely convincing images in test conditions, generally of well-known buildings, which he was allegedly picturing in his mind. If these findings are genuine, a study of the interplay between the thoughtographer and his "gismo" (the device Serios used for his experiments) might contribute to our understanding of what SLIders are doing, but more thoughtographers are needed.

SLI and PK

SLI is clearly closely related to PK, and we could be forgiven for speculating that it is a form of it. This may be so, but there are two very obvious differences.

Firstly, the evidence for PK, as observed in laboratory tests, involves very small objects, which the subject "wills" to move across a level surface. SLI, on the contrary, occurs in public places, and the object is relatively large. It does not follow that SLI requires a greater force than laboratory-PK, since it may involve no more than interrupting an electric current or operating a switch. But the electric girls manifestly exert a much greater power.

Secondly, PK, once it is hauled into the laboratory, becomes a matter of the subject making purposeful use of his will power, whereas SLI is almost always spontaneous and involuntary. We have seen that SLI rarely occurs under those conditions – indeed, we have learned to expect the precise opposite, a refusal to perform to order.

So, despite a superficial similarity, we have to conclude that SLI and PK are two very different kinds of event. Possibly the same power is used in both, but if so, it is used in a very different way.

Poltergeist phenomena (Random Spontaneous Psychokinesis)

To say of an activity that it is a random and spontaneous variety of something that is being investigated but remains unexplained, despite possessing a name derived from the Ancient Greek word for energy, does not tell us very much about it. It was really better off with its old name of "noisy spirit" – which is what the poltergeist was originally supposed to be and which may ultimately prove to be a more accurate label than the one scientists have bestowed upon it.

What, for example, possessed poltergeist victim Annemarie Schneider? Whether she should be classed as an "electric girl" is debatable, but the incidents associated with her certainly suggest as much, while demonstrating considerable resemblance to SLI. They began in November of 1967, when strange happenings started to occur in a lawyer's office in the small town of Rosenheim in Bavaria, Germany, where 19-year-old Annemarie was employed as a clerk. Neon ceiling lights and electronic appliances were most frequently affected. Fuses blew with no apparent cause. The telephone time announcement facility was dialed four or five times a minute of their own accord, sometimes simultaneously on four lines, something that to do normally would require very sophisticated equipment. Investigation by the eminent psychical researcher Hans Bender showed that the incidents occurred only during office hours, and only when Annemarie was at work; when she was on holiday or sick leave, the office was undisturbed. When she walked through the office, light fixtures would swing above her and bulbs explode even though they were switched off. These happenings were recorded on video. It wasn't only electrical appliances; pictures hanging on the walls were rotated, and one was videotaped turning through 320 degrees. Pages flew off the calendar. As in Angélique's case, wooden furniture was affected; drawers slid out of desks, a massive cabinet weighing nearly 400 pounds was moved twice.

It was clear that Annemarie was somehow responsible, but there was no question of her doing anything deliberately. Many of the incidents could not have been done by human means. She herself was much affected, sobbing and even screaming as mayhem surged around her. Psychological tests showed her to be a highly stressed person, and things became worse when her fiancé broke off their engagement. She was dismissed from her post and from several other jobs she took subsequently.[14]

No conclusion was ever reached, and the affair was simply classed as an instance of poltergeist activity. No parallel was drawn with Angélique or the Smyrna Girls, but reviewing her case now, it is evident that they were very similar, and similar, too, to SLI. Both are spontaneous and seemingly not deliberately willed by the subject; both manifest in everyday situations, rather than a laboratory.

Poltergeist activity takes countless forms and has a long history. We have reports dating from before the Common Era.[15] We have a detailed account of poltergeist activity in Maçon, France, from 1612.[16] Major Moor, owner of Bealings House, England, kept a detailed record of the ringing of bells in empty rooms in his Suffolk home in 1834, and the fruitless investigation that was carried out. California investigator D. Scott Rogo was inside an affected house when it was bombarded with stones hot to the touch,[17] and South African anthropologist B. J. F. Laubscher encountered the same phenomenon in a native context at Newouldt.[18] Although certain types of event recur – throwing stones, moving furniture, displacing household items are three frequent examples – each case is unique and presents a character of its own. The incidents cluster round an individual who is the focus, albeit unconsciously. We do not find this with SLI.

SLI and poltergeists

It would be neat and tidy if we could shoehorn SLI into the poltergeist box, but it simply can't be done. True, there are obvious similarities, in that both phenomena involve interaction between people and material objects, and that those people appear to be the source of energy that supplies the power for "impossible" physical events. But there are also profound differences. The Rosenheim case displays a feature that seems to be true of all poltergeist cases: the focus on the individual.[19] It is generally a fairly easy matter to relate the goings-on to a member of the family or someone else intimately concerned, who is liable to be an adolescent passing through a traumatic phase. This may be true of SLI, in some degree, for we have noted that the SLIder's state of mind is generally an important factor. But whereas Annemarie's temperament, attitude to her work, and private life perhaps provided a psychological explanation, this does not seem to be true of SLIders, who show no sign of using SLI as an outlet for hidden resentments or frustrations. Any SLIder could do anyone else's SLI, whereas poltergeist events are custom-made for the individual. In poltergeist outbreaks, everything is centered on the individual, and personal motivation is not hard to find. Though SLIders are in some sort selected, in that some people do it and others don't, the personal element is lacking.

Poltergeists and SLI differ, too, in the way they happen, and in their scale. SLI happens in the public street; poltergeists prefer to operate in the privacy of the subject's own home or workplace. All that happens in SLI is that a light turns on or off, and even if this happens frequently over a space of years, it's a relatively contained affair. Poltergeist outbreaks, by contrast, can involve considerable force, notably when massive furniture or other objects are moved, or when a house is bombarded with rocks. Heated rocks, at that. Such features indicate operations on a far more elaborate scale than simply turning a light out.

Machine-busters

The Austrian theoretical physicist Wolfgang Pauli was known among his colleagues for the negative effect he had on laboratory equipment, so much so that his friend Otto Stern banned him from his Hamburg laboratory. The "Pauli Effect" became notorious among scientists, and the story is related that on one occasion James Franck was performing some experiments at Gottingen when a costly measuring device failed to function. The incident was reported to Pauli with the comment that he couldn't be blamed because he wasn't there. However, it turned out that at that time Pauli was traveling from Copenhagen to Zurich and had a brief stop at Gottingen.[20]

We have already seen that a number of SLIders do a little machine busting on the side and have concluded that they are using the same powers in both activities. It is tempting to look for psychological motives – as with the Rosenheim poltergeist and even the case of Angélique, the individual could be unconsciously demonstrating her resentment at her working or social situation. Just how this translates into a material process could be seen in psychoanalytic terms as symbolic – in the same way that Pauli's effect on instruments was read as an expression of his resentment at being cast as a *theoretical* physicist rather than a practical one – a way of saying "If I can't *use* apparatus, at least I can prevent it being used by others."

As their label indicates, machine-busters are concerned predominately with material objects. Whatever the individual's reason for taking it out on office or domestic appliances, there is something about their bodies that causes the effect, irrespective of whether some psychological factor is also involved. For example, the Australian bank employee [141] who had continual trouble with appliances does not seem to have been even subconsciously motivated by stressful factors, since it manifested over a period of years and in a variety of situations.

The Edinburgh University's work in this field, under the late Robert Morris, is notable, if only because it is one of the few academic attempts to grapple with problems of this sort. It would surely be a reassurance, if not a comfort, to machine-busters if more were known about what makes them that way, and our study would certainly benefit. But as far as I know the project reached few, if any, useful conclusions.

SLI and Machine-Busters

In a sense, every SLIder is a machine-buster, though only on a very modest scale. The machines that he busts are relatively small, and the busting is usually limited to interruption of its operation. The effect is short-lived, and there is no permanent damage. In the great majority of SLI incidents the light is self-restored to normal. Interestingly, this seems also to be true of supermarket checkouts and the like, including elevators. When computers and other electronic devices in an office or bank are concerned, however, the maintenance staff frequently has to be summoned to sort matters out. Why this difference, which almost suggests that the perpetrator of the SLI is concerned not to be a nuisance to the general public?

Despite this difference however, the fact that many SLIders are also machine-busters encourages us to theorize that the same force, or one closely related, is being employed.

Ball lightning and other Balls of Light (BOLs)

Apart from the essential health-giving properties of light, some lights appear to be positively concerned for our welfare, and act to help us. You find it hard to believe? Of course you do. But what else can you make of this, reported by a Czech climber:

> My husband and I are keen mountain climbers. One morning in 1977 we set out to climb the highest mountain in Czechoslovakia, Mount Snezka. It was a sunny day, but halfway to the peak the weather suddenly turned bad and it began to snow. Although I know the mountain well, after an hour or so it was clear that we were lost. By 3pm we were exhausted and desperate. We were now in the thick of a gale and snowstorm. Then suddenly I looked to the left and saw a big blue ball, which was very near me and shone with a clear and warm light. I was very surprised, and I must admit I was also very frightened. I shut my eyes, but when I opened them the ball was still there. I asked my husband if he could see it, but he could not. Then, when I stepped forward, the ball moved with me, then it turned to the right and moved slowly away from me. I followed it, as if hypnotized by it. My husband asked why I had changed direction and said it was the wrong way, but he followed me although he still couldn't see the ball. All the way, the ball showed us the right direction and after two hours we arrived at the town. As soon as I could see the first houses, the ball disappeared. I am convinced that it saved our lives. I cannot explain this event, but it is true.[21]

Anyone could be forgiven for doubting that story, but a very similar account was sent to me, quite independently, by the German officer's widow it happened to while climbing alone in the Bavarian Alps:

> You will understand that this is rather a heavy mountain tour, but there is a good way as well up as down, but one must not miss it as I did. Having started a little late for the return, and light beginning to fade, all of a sudden I found myself in a really dangerous position. As a matter of fact one year later a young girl fell to death exactly on the spot where I realized myself to be in an almost hopeless position. All of a sudden I noticed a sort of big ball of light, and this condensed to the shape of a tall, rather Chinese looking gentleman. Extraordinarily I was not a bit frightened, and also not astonished, it all seemed then quite natural to me. The gentleman bowed, spoke a few words, led me a small path to the tourists' way, and disappeared as a ball of light. [22]

Ah, that's an easy one, we might be tempted to say; we don't have to think that the Buddhist religion maintains a mountain-rescue service on standby in perilous locations. Frau Falk's subconscious mind deserves the credit, materializing as a traditional Buddhist entity – in whose existence she doubtless believed – to reassure her. As for knowing the correct route down the mountain, evidently she knew this all along, but had forgotten... Well, yes, such an explanation has a plausible ring, but it doesn't entirely convince. What do we really mean when we theorize that her subconscious mind "knew it all along"? Had she subconsciously taken in the information on some previous outing? Does one forget something so important? And why the roundabout game with the BOL that turns into a bodhisattva and back again?

In any case, no such explanation will do for this third case, reported in 1899 by the traveler J. Shepley Part from West Africa:

> Shortly before leaving the forest country, we were benighted one night, and our guide did not know

> how far we were from the town to which we were bound. We had been marching for some hours in the dark along one of the forest roads when I and others saw what we took to be a lantern in the thick bush. It was peculiar, as it moved as fast as we did. It presently came out on the track, and an attempt was made to capture the bearer, which failed, and Mr. Ferguson [a native guide reputed to possess psychic powers] said we had better leave it alone. The impression I retain is as a focus of light throwing a circle of light round it, much as an ordinary stable lantern would.... As we approached, the light eluded us, and then followed the path ahead of us for some miles, and then disappeared, just outside the town we were approaching; it moved exactly as if carried by a man. The explanation given to me was the "double"... sent out along the route to act as our guide. [23] [27]

Even three such cases don't prove that anything more than hallucination is involved, especially as in the Czech case the husband does not see the BOL, which we must suppose would have been sufficiently visible for both to see it. But just as with SLI – where we ask "Why a streetlight? – here we ask, why in each case a ball of light? Is it because a ball of light is a kind of archetype, created by the traveler's subconscious mind simply because it is a kind of universal symbol?

The crux of the puzzle in all three cases is, of course, that the BOL "knew" what it was doing. It knew, first, that here were some humans who wanted desperately to get to a specified location, and it "knew" how to guide them there. Could this information have come to the BOL via the minds of the travelers? Yet all three insisted that they did not know the route. Or, if they did not actually know it, being experienced outdoor people, they might conceivably have worked it out for themselves. But in that case, why this game playing with BOLs?

Some years ago I presented a paper at a Ball Lightning conference at Hessdalen, Norway, under the title "Ball lightning in the wider context," in which I brought together various reports of anomalies and contradictions in reports attributed to "ball lightning." One substantial group of these comprised reports in which the phenomena – supposedly natural events involving natural forces – seemed to be closely associated with humans, generally the person who witnessed them. Though I don't question that they are indeed perfectly natural phenomena, the human tie-in invites comparison with SLI.

Terrestrial BOLs

> A very intelligent and well-balanced girl I know told me how, one night as she was walking home only an hour or so after dark, she saw a Will-o'-the-wisp [the popular name in the British Isles] moving slowly along the bog parallel with the road and not 50m away. As soon as she stopped to look at it, it suddenly changed course and began to come straight towards her, increasing its speed. This rather shook her determination to examine it with scientific coolness, so she hastily moved down the road a bit. To her consternation it changed direction too and still came straight for her, now crossing the soft bog as fast as a man could run on a road. This proved to be too great a strain on her morale and she promptly turned and made for home with all the speed she could. To this day she is convinced that it was directed by a real intelligence. [24]

This anecdote, one of thousands which tell a more-or-less similar story, includes the characteristic observation by the witness that the marsh light, will-o'-the-wisp, or *feu follet* – as the French call it – seemed to be behaving intelligently. Unlike the benevolent BOLs we have considered, however, marsh lights are generally reported as malevolent, luring the traveler into swamps or quicksand with the hope that where there are lights there is surely safe ground. Picturesque folklore? It's tempting to think so, but many accounts are well accredited and suggest that there is some substance to the belief. Even if we discount them all as subjective impressions, there remains the fact that the behavior and often the duration of the phenomenon defy most of the scientific explanations. When there was an outbreak of "fireballs" near Saratoga, Texas, in 1961, Edwin Hays, head of the biology department of the nearby Lamar State College of Technology, offered the "conventional" explanation of marsh lights – "swamp gas" – ignited by spontaneous combustion. Curtis Fuller, the admirably hard-headed editor of *Fate* magazine, responded:

> We would like to challenge Dr. Hays and every other proponent of this old chestnut of a theory to prove it. A common scientific criticism of psychic phenomena is that it is not repeatable under laboratory conditions. Well, we challenge the swamp gas theorists to produce a little swamp gas under laboratory conditions, ignite it spontaneously, and produce their own little fireballs that go bouncing merrily along for minutes at a time. [25]

There is more than one difficulty with the marsh-gas explanation, but the most serious objection is that of duration. *Feux follets* have been observed for periods up to one hour, sometimes even more. If they had remained in the same location, we could hypothesize that they were sustained by a steady supply of fuel from that particular portion of the terrain – a natural source of methane gas or some such. But this is not the case. On occasion, investigators have followed them as they moved in a seemingly purposeful manner, over a distance of a kilometer or more, as in the following account:

> In October 1938 I had developed an interest in the folklore of the Ozarks. I had been told that if I went to a certain old sawmill pond after dark I would stand a chance of seeing one of the local spooks – a ball of blue light rising from the ground and drifting into the woods. With the fellow who had told us about it my dad and I went to the millpond one evening carrying an old lighted barn lantern. We had been there for maybe 20 minutes when around the end of the pond about 10m away, a small blue light began to form near the ground. It was about the size of a softball and its color was a clear cornflower blue. Dad picked up the lantern and we started toward the strange light. By the time we were 3m from it, it was about 1m off the ground and had expanded to the size of a basketball. Its color was still clear and bright but now more a robin's egg blue. We were less than 2m from the light when it started to move away. Our companion picked up a piece of a tree limb and shoved it into the ground to mark where the light had first appeared.
>
> We started following it down the road. It had now taken up a swinging motion: the high point of the arc was about 1.3m above the ground and the low maybe 0.2m. At its lowest point it gave enough light for us to see the white sand of the road. The light went south about 200m to a one-room country schoolhouse. There it left the road, moved over to the building and passed from front to rear along the north side. We followed it along

> the back side of the school and back toward the road where it started south again. This time it went to a fork in the road, took the left hand road and went about 600m to the gate on a lane leading to an abandoned two-room log house.
>
> The light went over the gate and by the time we opened it and passed through, the light had reached the house. We watched it make a complete trip around the house and then it went inside through the only door. We could see it passing the windows and through the cracks between the logs. It came out after a few minutes and started back toward the road. At this time the light passed within 0.3m of me, moving smoothly about 1m above the ground. It emitted a low soft hum like the 60-cycle hum of an electrical transformer.
>
> The light took pretty much the same route back to the pond without the detour around the schoolhouse. We were 6-7m from it when it reached the tree limb sticking out of the ground. Hovering there for perhaps half a minute, it became smaller and darker in size. When it had reduced to the size it was when we first saw it, it settled into the sand and was gone.
>
> I don't think this was ball lightning nor was it the classic will-o'-the-wisp. To this day I have no idea what it was – but three of us saw it and followed it for most of an hour. And it did return to its point of origin. [26]

No natural fire has the capability to sustain itself over so long a period, let alone travel over so long a distance. The return to its place of origin is a feature that the narrator is right to emphasize, for this is particularly hard to explain in terms of any natural phenomenon. Even if we set aside the behavioral characteristics of the object, its physical properties are sufficiently baffling. Moreover, though some details suggest that this was a *feu follet* of some kind, no convincing explanation has ever been given, as Fuller justly points out, as to why marsh gas should spontaneously ignite in the first instance.

Ball lightning

Even though the existence of ball lightning (BL), as a phenomenon, remains open to question, and its nature even more so, it is a convenient term for a luminous sphere, generally gaseous in substance, generally a meter or so in diameter, which generally manifests in association with storms. So many "generally" warns us that there are many exceptions to the typical specification. It has attracted the interest of scientists worldwide, notably in Japan and Austria, who present many different theories as to what ball lightning is, and even whether it exists at all as a phenomenon in its own right.

That ball lightning exists as some kind of phenomenon is not in dispute, however. It is what ball lightning *does* that challenges the classification. Here are two cases from 19th century France that exhibit the astonishing behavior of which it is capable:

> *November 1898*: A fireball appeared in a room where a girl was sitting, crossed the floor towards her, circled round her in a spiral, then darted up the chimney and exploded out of doors.
>
> *August 1895*: A fireball passed through a farmhouse room, ignoring two people, passed through the floor (making a hole as it went) into the sheep-fold below where it killed five sheep (though with no sign of burn or wound) yet left unharmed the sheep-boy who was with them.[27]

We simply don't know whether BL likes people or not. Clearly, in those two cases, the BOL was quite capable of killing the humans it encountered, which implies, if not benevolence, then tolerance at the least. Once again, though, we are denied the chance to establish a positive across-the-board parameter for the phenomenon. For while it is true that BL *rarely* harms people, it is also true that witnesses *have* been injured by BL. So the most we can say is that BL shows a *curiosity* about humans, pretty much as a kitten might do, and perhaps we might go so far as to say that it shows a reluctance to harm humans. But even this is more than an unfeeling force of nature can be expected to do.

Ball lightning is not a kitten, nor any kind of animate thing. It is a natural phenomenon, born of other natural phenomena. Yet many hundreds of narratives are on record, chronicling such details as its ability to pass through solid matter. Often it does as in the second case cited above, but generally it uses existing routes, as in the first case where it uses the chimney.

SLI and Ball Lightning

Because BL is so poorly understood, we cannot draw any useful conclusion from a comparison. But the very fact that a comparison suggests itself shows that there may be lessons we can learn. From scores of such narratives as those we have cited, it is conceivable that BL has an interest or concern about the humans it encounters, and as with SLI, its behavior is affected when it meets you or me.

Orbs

A relatively recent form of light manifestation, orbs have been derided by skeptics, but proclaimed by their champions as a genuine phenomenon. They generally consist of small BOLs that appear in digital photographs. This circumstance encourages the view that they are optical artifacts with no significance, and this is how they are widely regarded.

Others, on the other hand, believe them to be meaningful. They assert, for instance, that orbs are likely to show up when a healer is photographed at work. In their view, orbs are spirits or living creatures of a sort, invisible to the naked human eye but captured by the wonder of digital photography. The evidence is weak and will remain so until an orb is captured on film giving unmistakable evidence of purposeful activity of some kind. Nevertheless, the similarity of the orb phenomenon to our puzzling BOL cases, and the allegations of involvement with mediums and healers, warn us that this curious development should not be dismissed outright.

Religious BOLs

Lights are mentioned in countless narratives of saints and religious figures, and in association with religious occasions and sites. The most thoroughly documented are the Egryn Lights, which manifested during a religious revival in Wales in 1905. What happened here was that low-altitude aerial lights were seen on several occasions by numerous observers, hovering either over chapels where a prominent preacher, Mrs. Mary Jones, was scheduled to appear, or over the dwellings of future converts. The conclusion that the lights were physically real, and that they were associated with Mary Jones, can be avoided only by denying the validity of the testimony, and that is not easy to do, considering the large number of independent witnesses, many of whom – notably the reporters from the national press and investigators for the Society for Psychical Research – were initially skeptical.

But if we accept the observations as real and substantially accurate, it is hard to see how we can avoid the logical consequence – that the synchronicity with the activities of Mary Jones was purposeful, not coincidence, and that the lights were somehow generated by Mary herself, or in association with her. This is so contrary to received experience that most of us would prefer to reject the testimony, however well confirmed. We could derive some comfort from the fact that there existed a long-standing tradition of anomalous lights observed in the area, due to the geological character of the terrain. It's a nice scientific explanation but it hardly explains the synchronicity with the timing and location of the religious events.

But if science can't help us, do we have to suppose a miracle? There is a third option, that Mary possessed some faculty which enabled her to interact with natural forces already existing in the area. To pursue such a line of thought may seem to be going beyond legitimate speculation, but the fact is, the phenomena challenge us to account for them *somehow.*

Mountain lights

Recurrent lights, seen generally in mountain locations, have been reported from many parts of the world and can safely be classed as natural phenomena though their nature and behavior often elude explanation. The Piedmont Lights, occurring in a mountainous region of Missouri, were the subject of an intensive investigation led by Harley Rutledge, president of the Missouri Academy of Science. Challenged by his students to explain this nearby phenomenon, his team recorded 178 anomalous objects on 157 occasions, with independent instrumental corroboration. That alone would make his "Project Identification" of the greatest interest, but there is more for our purpose: on at least 32 recorded occasions, he and/or his colleagues recorded a high degree of synchronicity between the movement of the object and the activities of the observers. These might be as physical as switching a car light on or off, or a verbal or a radio message between the investigators, or even an unspoken thought in the mind of one of the observers. Of course, there is no way of substantiating this third category; nevertheless, if we accept the investigators' word in other respects, we must at least give them a hearing in regard to these remarkable claims.[28]

It seems likely that what is involved in such cases, not only in psychical research but even in hard science, is a variant of the Experimenter Effect, which is beginning to impose itself as a secondary problem in so many fields of research. We need not go so far as to suppose that the human subject is responsible for actually *creating* the anomalous object; it is enough to suppose that some kind of interaction is taking place between the observer and the observed.

I don't suppose any of us feel very comfortable with this type of explanation. I certainly don't. Once we admit the implication of a human element, we seem to be opening the door to all kind of parapsychological and even occultist theorizing. But if we insist on a scientific explanation – and surely we do – then science must accept mutual interaction between people and things as something more substantial than folktale and legend – as a fact of life that merits investigation along with the rest.

SLI and BOLs

There would seem to be absolutely no connection between these various BOL phenomena and what SLIders do, apart from the feature that an ostensibly inanimate object seems to take an interest – usually a friendly one – in a human being, to the extent of understanding their predicament and assisting them out of it. But this is saying a great deal. In both cases, something is happening that is not supposed to happen. It is as though the object has a mind of its own. In the case of the BOL, it seems to sense the human's need. In the case of the streetlight, the light seems to sense the SLIder's state of mind, and then, to know whether or not the SLIder is *willing* it to extinguish.

So while what they do in either case is very different, both alike testify to a power in nature that we have not hitherto recognized, but, like the experimenter effect, must learn to take into account.

Lightning victims

A further manifestation suggesting some kind of relationship between people and supposedly inanimate phenomena is the fact that certain people are reported as being repeated victim of lightning strikes. This is normally attributed to mathematical chance, and so it may be. But we should at least entertain the possibility that some people are particularly liable, due to their physiological make-up, to this kind of accident.

Roy Cleveland Sullivan, a forest ranger in Virginia, survived being struck by lightning seven times during his 36-year career. He once made the remarkable statement that he could see the lightning traveling towards him.[29]

Okay, we might say that a forest ranger, by virtue of his occupation, is especially liable to such a risk, and perhaps Sullivan was working in a region where lightning is particularly common. But is the same true of all such accounts, which have been meticulously collected by the editors of *Fortean Times?*

Glowing nuns & other luminous people

The Roman Catholic Church lays down rules for everything, and one of them decrees who is entitled to be depicted with an aureole (= halo) and who is not.[30] But to have a golden circlet hovering over your head is not the only sign of sanctity; it shows itself in other enlightening ways. Throughout history there have been reports of luminous phenomena seen in association with religious notables. If reports are to be believed, these signs of divine grace had their origin in fact.

But are these reports to be believed? Many serious commentators, while critical of other pious legends, share the feeling that perhaps there may be some substance to the reports of luminous phenomena occurring in association with religious states of mind. The German historian Joseph von Görres devotes an entire chapter of his definitive work on Christian mysticism to religious figures who became luminous, wholly or partially, usually when they were in an ecstatic state:

> Colette van Gent [Flanders, 580-1646], was often completely illuminated while praying, to such a degree that more than once the sisters ran to her cell, supposing that it was on fire. Once her veil was found scorched, though there was no source of fire nearby.
>
> Saint Gertrude of Nivelles [Flanders, 626-

659] told her biographer that one day, while she was in prayer before the altar, she saw descend upon her a luminous globe which lit up the entire church. This lasted for a half-hour, then disappeared little by little.

Saint Leo of Catania [Sicily, 703-787] would often pray in the church with a lay brother of great piety. Throughout a long period, a peasant who lived nearby saw at night a globe of light rise from the roof of the church. One night he resolved to investigate and saw two beautiful lights rise from the church into the sky. More astonished than ever, he rang and woke the porter of the convent. This man, who knew him, opened the church door, and both saw, as they entered, Leo and his companion in ecstasy before the Holy Sacrament, raised up in the air.

Christine Mechthilde Tuschelin, [Germany, 13th century] of the convent of Adelhausen, was frequently enveloped entirely by a light so bright that no one could look at it directly. She was obliged to remain in her cell so that the sisters could go to the chapel...

Violante, queen of Aragon, one day saw her confessor, Saint Vincent Ferrer [Spain 1357-1419], surrounded by light while he was at prayer.[31]

Your instinctive reaction to such stories is probably the same is mine – that they are well-intentioned inventions intended to bolster the reputation of those concerned. At this distance in time it's impossible to say with certainty, but the sheer weight of testimony is impressive. Though they may seem to take us a long way from today's SLIders, they, too, seem to indicate the possession by individuals of some kind of internal force. Are they comparable to a more recent case, that of Anna Monaro, an Italian woman who in 1934, while lying sick in a convalescent home, gave off a bluish light from the region of her chest? She was thoroughly studied, but though her condition must be supposed medical rather than miraculous, it seems to have been a one-of-a-kind phenomenon.[32]

Perhaps Anna should not be classed along with the medieval glowing nuns and saints, but it is striking that the old storytellers should pick out this feature, which presumably is seen as symbolic of their sanctity. How did luminosity acquire this connotation?

SLI and luminous people

As with the other phenomena mentioned in this chapter, things like this can't just *happen*. Natural forces are involved. Both SLI and luminosity require a certain amount of energy, and that energy has to come from *somewhere*. But whether it's a glowing nun in 13th century Germany or a hospital patient in 20th century Italy, what other energy source is there except the individual's own body?

The power within

The human body can do a lot of strange things, and a lot more strange things can happen to the human body. Gould & Pyle's 900-page *Anomalies & curiosities of medicine* documents many of them, and researchers like Vincent Gaddis have explored specific topics. There we can read of people who shine in the dark or who are too hot to touch; of people like Angélique who attract or repel objects; individuals like the 19th century medium D. D. Home who could handle glowing coals or like indigenous people in many parts of the world who can walk on them with bare feet.

From Angélique Cottin to Anna Monaro, from Annemarie Schneider to the rescued climbers, these people's experiences serve as a reminder of the complexity of the human body, and of its interactions with the universe at large.

It took power for Angélique and the Smyrna Girls to repel furniture. It takes power for Uri Geller to bend a spoon, just as it takes power for a machine-buster to devastate a computer-system, for a house to be bombarded with stones that are hot to the touch, or for a patient's sickness to be alleviated by "spiritual energy." And, of course, to do SLI.

Everything that happens requires the deployment of energy in one form or another, but the activities we have been looking at in the course of this chapter have one further feature in common: the energy is *human* energy. We don't know how the energy of a poltergeist heats the stones before hurling them at the targeted house; we don't know how a computer-killer generates the force she needs to immobilize a bank's computer system. It may be, as we hinted earlier, that they simply pluck it out of the air. But somewhere, somehow, these people find the energy and find ways of using it for their purposes. Similarly we have hypothesized a SLI-force that is deployed by SLIders, and we have seen that although SLI-energy has much in common with other energy-requiring activities, there are also considerable differences. Of these, perhaps the most marked is that SLI does not call for will power; if anything, it requires that will power *not* be exerted. The same is true of other incidents: housewives do not consciously wish their washing machines to die on them, bank employees do not want their jobs endangered by crazed computers.

So drawing parallels with other strange happenings may not help us much *directly* in our quest. SLI *is* neither PK nor poltergeists at work. On the other hand, there *are* similarities; it is not coincidence that the SLIder from Somerset, mentioned earlier in this chapter, does SLI in addition to the non-SLI things she does. Of course there is a connection. What is happening in SLI is not the same as is going on in these other happenings, but the channeling of energy, in what appears to be a planned and purposeful activity, is common to them all So we will do well to continue to explore all of them, noting the differences as well as the similarities. For surely, this is where the solution to the SLI enigma will be found.

Chapter Nine

'Things are against us'

A SLIder from Wimbledon, London, UK, after describing his SLI experiences to a friend, says:

> I went into my bedroom to fetch the encyclopedia to show him the article describing the SLIDE Project, when the light bulb blew. [167]

Coincidence? No doubt – but how exquisitely timed! Naturally, every SLIder tries to rationalize his/her experience, and when straightforward explanations fail to satisfy, some have come up with more way-out possibilities. A SLIder from Harrogate, England, had an ingenious thought:

> I have a reasonably scientific mind and was trying to rationalize the situation, coming up with various ideas such as the council experimenting with movement detectors so that the lights would turn on as someone approached and off as they passed! I couldn't convince myself with this idea, and the other ideas were even less likely. Not being able to explain it, I was spooked every time. I told my friends and to silence the piss-takers I took them past the light post one dark evening, Sod's law, nothing happened, not even a flicker. [178]

Later he realized that the reason he couldn't demonstrate his ability to his friends was because he was no longer in the right frame of mind. When on his own, he was "slightly nervous (fear of being mugged, I suppose)," whereas with his friends he was in "a relaxed situation." But though he correctly identified the necessary conditions, that didn't help him reach a satisfactory explanation. In the previous chapter, we identified the interaction between mind and matter as the paramount feature that SLI shares with several other anomalies, but we are no farther towards understanding how that interaction comes about. For is there another mechanical action known to science that lays down conditions when it will do as the operator wishes, or operates only when the individual concerned is in the "right" frame of mind? Wolfgang Pauli would have relished the dilemma!

And yet, yes, maybe there is. Psychical researchers have become increasingly familiar with the "experimenter effect," where the outcome of an experiment or process is unwittingly and unconsciously affected by whoever's doing it.[33] In 1981 Rex Stanford published a paper in the *Journal of the American Society for Psychical Research* entitled "Are we shamans or scientists?" in which he presented the case for taking into account attitude and personality factors associated with individuals when evaluating the results they achieve. Later in this chapter we shall see some examples that seem to manifest the ability of a lifeless appliance to "read" the state of mind of someone working with it, or simply passing near it, and apparently to adjust its own behavior accordingly. Such an action flies in the face of all we think we know about how the universe operates, yet in the light of the experiences cited in the preceding chapter, it becomes easier to believe that some such influence may be at work and may be what is happening in SLI.

Our lives are rooted in the belief that there is a satisfactory solution to every problem. The solution may be evolution, or it may be God or the fairies or Fate or the Furies (whom the Greeks termed "the kindly ones" not to hurt their feelings). But somewhere, we believe, for every blank square in a crossword puzzle there is a letter waiting to be inserted. In the earlier chapters of this inquiry, we reached a working hypothesis of what is happening in this particular puzzle. Then, examining the human factor, we found indications that personal circumstances play a significant part. Comparing the SLI experience with other strange happenings, we found clear parallels supportive of the idea that some of us possess, even if only at certain times or under certain conditions, a power to inspire what has every appearance of cooperative action with inanimate things. But though this is helpful as far as it goes, it is clear that some element is still eluding us: the *why*, the reason underlying this seemingly unreasonable phenomenon.

Earlier in our inquiry, it seemed increasingly clear that the crucial element in the SLI enigma is the individual. Streetlights can't make choices or decisions; they surely can't play any active part in the process. But the phenomena we looked at in the preceding chapter make it less easy to answer with a confident "No, of course they can't." We saw supposedly inanimate objects playing – or at any rate seeming to play – an active role, apparently making choices and decisions of their own. So in this chapter we go back to things for a while, to see if we may have missed something.

Resistentialism

In 1950 writer Paul Jennings, whose column "Oddly enough" was a regular feature of the London Sunday *Observer*, wrote a disrespectful piece about "Resistentialism," a spoof on the currently fashionable French philosophy of Existentialism.[34] Though the intention was satirical, the piece struck a chord in readers who sympathized with the movement's slogan: *"Les choses sont contre nous"* – "Things are against us." He described, for example, an experiment allegedly conducted by French scientists, who dropped slices of buttered toast on the floor, finding that the more expensive the grade of carpet, the more likely the toast was to fall butter-side down.

Each of us has had momentary feelings of frustration with such activities as assembling flat-pack furniture, or rain on the day of the barbecue. I once had a flat tire while driving with a girlfriend to catch the Dover Ferry to France; you have surely had experiences just as vexing. As for the caprices of computers when they get into one of their bolshie moods, the least said the better. SLI can readily be seen as a symbolic confrontation between man and the devices he has, Frankenstein-like, brought into the world. These are the things we remember, forgetting the many times I drove to the Dover Ferry and *didn't* get a flat tire. Ah, but not with this particular girl.

However, it is no less true that there are people who seem to have an affinity with machinery. Motor mechanics who seem to be able intuitively to diagnose a fault, while others are adept with clocks and watches and can fix the TV with their eyes closed.

So *le résistentialisme* tells us more about ourselves than it does about SLI. But that's no bad thing. For if it is abundantly clear that the human factor is fundamental to understanding SLI, we should make allowance for other factors. Among the Azande people of Africa, if a woman fell sick or a man's house burnt down, it was assumed that an ill wisher had hired a witch to inflict the misfortune. Like Freud, but with a quite different rationale, they took it for granted that "there are no accidents." The Azande would not hesitate to tell you what causes you to cause SLI: some fellow has put a jinx on you.[35]

The JOTT phenomenon

Thought provoking, too, is the pioneering work of lawyer Mary Rose Barrington, a Council Member of the British Society for Psychical Research, on JOTT ("Just One Of Those Things"). Formally defined as Spatial Discontinuity Manifestations, these are the seemingly casual happenings in which objects – usually personal possessions of no great value – appear, disappear, and reappear in circumstances that seem at least improbable, more often inexplicable, or behave in other ways incompatible with scientific explanations. Because of their trivial nature, it is easy to dismiss these incidents as coincidence or misperception. But a number of instances have been investigated so thoroughly that no explanation other than Azande magic seems to makes sense.

Barrington distinguishes between six types of jott. The most common form is the *jottle*, which is the displacement of an item, though not as in a poltergeist outbreak where objects are typically thrown around in full view. Instead, *jottle* takes place surreptitiously and is rarely observed happening. It's simply that the screwdriver you just placed at your elbow somehow isn't there when you reach for it. As Barrington described it to me:

> The most basic form of jottle is when an article apparently ceases to be in the place where it was definitely known to be at an earlier time, sometimes a matter of minutes, or moments, ago. Sometimes the article is later found back where it ought to have been all the time, sometimes it is never seen again, sometimes, most interestingly, it is found in a different location, one that it could not have occupied in the interim, and often a place neither you nor anyone else would conceivably have put it.

Barrington has collected examples of the various forms of *jottle*. The *comeback* is the item that comes back to the place where it was before it disappeared, or to a "place close by." It is easy for the loser to think she has been mistaken, or is the victim of a hallucination. A variant is the *walkabout*, the item that reappears, or is found, in some other place, often an improbable or incongruous one.

A *flyaway* is the article that vanishes seemingly forever. I had a large kitchen spoon that I used every day, which vanished in this way. Too big to be tidied away in a drawer or dropped into a waste bin, yet gone, never to be seen again.

But the most bizarre is the *trade-in*, where a very similar article appears in place of the one that flew away, as if trying to pass itself off as the one that went missing. The *trade-ins* are not only the most extraordinary, but also the strongest evidence that an intelligence of some sort is responsible. In one of Barrington's cases, for example, a woman "lost" a brooch that she wore on her outdoor coat. Sometime later she was pleased to see it back again, pinned on a dress; yet it wasn't the same brooch, but one which strongly resembled it. Apart from the mystery of the disappearance and reappearance, we wonder where the replacement article came from. Do we picture the mischievous entity searching the stores till it finds a near match and saying hopefully to itself, "Oh, she'll never notice the difference"?

Barrington sets out the nature of the puzzle:

> We are talking about articles that cease to be constituted as identifiable items at one location and are reconstituted, in the case of walkabout, at another location. I say identifiable items, because the article can also be described objectively as an area of mostly empty space in which there is an ordered assembly of particles or electrical charges or oscillations, and which acquires a certain character – as, say, a hairbrush or a ring – by the observer.

When an item is the subject of a jott, it vanishes – that is, something is done to its physical structure. Jott differs from SLI in that no human force appears to be involved, but rather the opposite. On the other hand, perhaps the owner, or someone, provides the energy whereby the jotting takes place?

Barrington speculates:

> So to conclude, jott suggests an idealist sort of cosmos, though one that works on and with the physical raw material that constitutes the familiar world around us, the world explored, exploited and explained by science. Though this domain

> appears to be governed by immutable laws these can sometimes break down, or can be broken down, giving us glimpses of this different level of reality.

Barrington's conclusion seems to be one of despair at finding an explanation of her vanishing items on our existing, everyday level of reality. Shall we be forced into the same impasse in our investigation of SLI? Jotts and SLI are not the only happenings that have driven investigators to conclude that it's no use looking for an explanation in terms of our present view of reality, that we need to broaden our mental vision to include other possibilities.

Switching on

A trivial happening we have all experienced is finding an appliance switched on which we don't recall switching on – a typical jott situation. A Scottish SLIder told us:

> At that time I was working in TV and was a particularly stressed individual. In the early hours of the morning a buzzing sound woke me and I looked up to see a small portable TV sitting on a chair in the corner of the room was switched on. It showed only noise and was not tuned to any channel. My bedclothes were undisturbed which made me feel that I could not have got up and switched it on and the creaking floorboards of some other person would have woken me before the buzzing sound. In itself this would not be too unusual but when I put on the light and went to switch the TV off I discovered it was unplugged. I didn't panic. I went back to bed. Later, I was woken again. This time I was sweating with the heat. The electric blanket, which was most definitely switched off, was now switched on to

> full... I was stressed at the time and believe this might have something to do with it. [198]

Coincidentally, about the time I received this email, I myself had a similar experience. In the early morning I woke and heard a buzzing sound; the house alarm, which needs a code – known to no one else in the house at that time – to switch it on, was sounding. Also the central heating was switched on, though it never is at night. It happened that this night – and no other night before or since – a woman, with whom I had no personal connection but who was likely in a stressed state, was staying in the house. She would not have been able to set the appliances herself, yet I have little doubt that it was her presence in the house that switched on the appliances.

Perhaps a poltergeist was responsible. Perhaps the woman was an Electric Girl.

Séance-room phenomena

The curious things that happen at spiritualist séance-rooms come about because people are hoping to communicate with the dead, and they believe that specially gifted people – *mediums* – can help them do this. This belief has always been exceedingly controversial, and though many eminent scientists have espoused some form of spiritualist belief-system, the scientific world in general is as skeptical today as it was when spiritualism was launched, nearly two centuries ago. So to bring in séance phenomena to help us understand SLI may seem like invoking one mystery to explain another.

However, though we may disagree as to how these phenomena are interpreted, there is no question but that some very strange events take place in the séance room or other places where spiritualism is practiced. Often, they involve the manipulation of physical objects, perhaps in the same way as PK, which we glanced at in the previous chapter, and SLI. For example, Glen Hamilton, a respected Canadian physician, carried out many experiments with spirit mediums. A feature of these séances was the ringing of a bell in a box beyond the medium's reach; the circuit that caused the bell to ring could only be closed by closing the lid of the box, which was held open by a spring that required a force of 400 grams. It was alleged that Walter, the medium's control, did this via an ectoplasmic cord leading from the medium's mouth.[36]

This is just one of thousands of reports in which physical incidents – furniture raised in the air, articles moving in sealed containers, that kind of stuff – have allegedly taken place, often under strict control, though nearly always in darkness. Despite countless hours of experimentation, resulting in some very impressive findings, no incontrovertible proof has been obtained. Many believe that spirit controls, who are supposed to be the link between the medium and the departed, are not and never have been real people, but secondary personalities of the mediums. Whether or not this is so, it is plausible that some entity – be it a discarnate spirit or an extension of the medium's personality – is causing the phenomena. This is not so very different from what seems to be happening in SLI, but there is not the slightest evidence that it is a secondary personality of the SLIder that is manipulating the lights. This does not rule out the possibility that the lights are controlled by spirits of the dead, or even of the living for that matter, but no one has seriously suggested it. It's just one more possible line of investigation.

The paranormal

The general feeling among investigators is that SLI is a natural event for which there is currently no normal explanation. This would make it, by definition, abnormal or paranormal. However, this is not to say that it is supernatural. Like poltergeists and "electric girls," the explanation surely lies within the scope of science, though it must necessarily be a more extended science than is currently available. Three California SLIders make an explicitly paranormal connection:

> *Carlsbad:* I have "practiced" witchcraft in my youth, mostly harmless influencing people. I feel in close harmony with the Earth and many creatures, particularly dolphins and cats. Once I had a black cat and when I had a fatty cyst show up on my neck one also appeared on his at the same time. They both went away at the same time. Both my father and my brother had/have the same effect on watches, but I don't know about streetlights. [27]

> *Los Angeles:* As a child I was quite psychic. I gradually "shut down" over the years and am reopening, seemingly as a result of my metaphysical/spiritual disciplines. [34]

> *Rohnert Park:* I chant and meditate daily and I think this has expanded my consciousness a little bit to be more "electric" or sensitive to the magnetic and perhaps other fields of energy that comprise ourselves and our environment. I haven't always been "electric." [35]

And a radio announcer from Pardeeville, Wisconsin says:

> I often have precognitive dreams about things that will occur the very next day... My ashtray jumped up and split in two... My lover saw my spirit sit up while I lay sleeping. [54]

What do these reports tell us? That these people are kooks? Maybe, but we have no right to dismiss their experiences as irrelevant to our quest. After all, those who have no first-hand experience of SLI tend to think of SLIders as pretty kooky anyway. SLIders must accept the fact that, provisionally at least, they are classed along with machine-busters, weather-prone people, healers, mediums, and others whose faculties take them to the outer fringes of human existence.

Ultimately, the evidence that things are against us is inconclusive; on the contrary, it is well nigh certain that things don't give a damn about us. Skeptical Shropshire poet, A. E. Housman, was nearer the mark when he wrote:

> ...Nature, heartless, witless nature
> will neither care nor know
> whose feet they are that tread the meadow
> and trespass there and go,
> nor ask, amid the dews of morning,
> if they are mine or no. [37]

Except that it isn't always so, is it? What about those lights that guide travelers in need of help? Not everything is against us.

Chapter Ten

Making sense of SLI

Exeter, England: At first I assumed everybody else was getting caught out and imagining themselves latter-day Uri Gellers, but I soon rationalized that were this the case, then the streets of Exeter and its suburbs would be flashing on and off like Christmas trees. [175]

San Francisco: If everyone saw as many lights going out as I do, then there would soon be none on at all. [111]

Australia: At first I thought it was just coincidence, but simple calculation suggested that if the lights went out at this rate for everyone there would be no operational lights at all. [112]

A good many SLIders have uttered words to this effect. Is it a real danger? How many SLIders are there? Not enough, it seems, to justify imposing a curfew on known offenders, requiring them during the hours of darkness to stay indoors where they can wreak less havoc.

Most people don't know about SLI. The SLIders themselves may tell a few friends, but don't otherwise broadcast it. They don't see it as something to brag about. There have been a few press articles about our investigation, as well as talks and interviews on radio or TV, but usually early in the morning or late in the evening, when they could have reached only a tiny fraction of the population. And even those who know about our investigation don't necessarily take the trouble to get in touch with us, or even know how to do so. So the 200 plus people who have contacted us can be only the tip of the iceberg. If we haven't heard from SLIders in Uruguay or Uzbekistan, it's not because there aren't any, but because they don't know of our activities. There must be thousands, possibly millions of SLIders around the world who have not heard of us and probably think, as so many of those who *have* contacted us used to think, that they are suffering from some rare and possibly unique affliction. But what would happen if, perhaps as a result of this book, many more SLIders realized their potential? Rallying cries – "SLIders of the world unite!" – might bring them together to use their potential to devastating effect. We were amused by a letter from a Canadian professor who told us:

> I have been turning off streetlights for years, but I have always hesitated to discuss it because it makes me sound like a loony. Last year, while teaching at a university in the United States, I became friends with another professor who mentioned to me one day that she turned off streetlights. Needless to say, she was surprised when I was not taken aback! We had an interesting time comparing notes. [159]

Imagine such incidents on a larger scale. So far, SLIders have shown themselves well intentioned, unaware of the power they possess. But if they chose to join forces – literally – what then? Why, there's all the makings of a disaster movie ready and waiting…

As things are, SLI is a rogue force, a loose cannon, neither understood nor controlled. But how can we learn more? A few organizations have contemplated carrying out formal investigation into the phenomenon. The Research Committee of the London Society for Psychical Research, under the late Arthur Ellison, and the Psychology Department of the University of Hertfordshire, under Richard Wiseman, have both looked into the possibility of mounting an investigation. But both projects came to nothing due to the seemingly insurmountable difficulty of finding subjects who can do SLI to order and replicate it so they and their talents may be studied. This is disappointing because a professional analysis of SLI and what they do would surely tell us something useful about human behavior, of potential value to both psychologists and parapsychologists, not only about SLI but about the human mind in general. At Edinburgh University, Scotland, a team led by the late Richard Morris tackled the puzzle of machine-busters and the like, but there was no head-on study specifically of SLI.

In addition there have been a number of television documentaries and interviews featuring the phenomenon, notably from Japan and from the Discovery Channel in the U.S. But they too have failed to find a subject who can turn off a light as easily as the rest of us switch off the kitchen light.

However, there is no need to be discouraged by this lack of progress. What we have seen in this book is encouraging for the future. First and foremost, the sheer weight of responsible testimony means that there can no longer be any doubt that SLI does really and truly occur, and that it is a genuine phenomenon not like any other.

Second, whatever SLI may turn out to be, it is clearly the result of an interaction between people and physical, man-made objects. This is territory largely unexplored by science, though parapsychologists have been investigating events on this particular frontier for a century or more. The stumbling block has been, and remains, our ignorance of the human mind and what it can do.

Questions and tests

SLI is a scientific enigma, but which discipline of science is the one whose job it is to tell us what is going on? The probable answer is: several disciplines working together. The engineer, to tell us what is going on inside the lights, and the psychologist to tell us what is going on inside the people who affect them. But before they are done, they will perhaps need to enlist the services of the anthropologist and the parapsychologist who have direct, first-hand experience of anomalous behavior. The hard scientists will do well to recognize that they can benefit from the experience of those who have studied magic and folk belief in indigenous cultures, and of psychical researchers who have done their best to elucidate anomalous behavior.

What we would dearly love to find is a SLIder who can perform to order. We have seen that many SLIders do SLI deliberately on occasion, either to satisfy their curiosity or to demonstrate it to friends. So it is a reasonable expectation that sooner or later we will find a SLIder who can do it deliberately – perhaps not every time she tries, but sufficiently often for her to be tested. Parapsychologists have discovered subjects who can demonstrate other skills, such as PK, as well as card guessing and remote viewing, for example, so we have a right to think that perhaps some day an investigatable SLIder will be found.

If and when that happens, what are the questions investigators will want to ask? Here are some that arise from the reports in the preceding chapters; doubtless there are many more, for the fact is that *any* scrap of information may help our understanding.

About the lights

Are the lights that SLIders affect invariably controlled either as a group or independently?

What precisely is being done to the lights?

Is there any discernible difference between the lights affected and those that are not? Light(s) that have been affected by SLIders should be examined to see if they differ in any way from the norm.

Is the distance between lights a relevant factor – i.e. must the SLIder leave the vicinity of one light before she moves within SLIding distance of the next?

More study needs to be made of the different ways in which

the lights respond to SLIders. Which types come on again automatically and more or less immediately, which after an interval, which only the following evening?

About the SLI-force

How strong is the force – if force it is – that the SLIder is giving out? What is its nature?

Over what distance can it be detected?

Is it in any way directional, or does it take the form of a force field?

How long does the effect last? For instance, what happens if a SLIder first extinguishes a light, then sits down underneath it? Will the light come on again – i.e. is it a short, sharp burst of energy? Or will it stay out as long as the SLIder sticks around?

About the SLIder

What does a brain scan tell us about the SLIder when she is in and out of the SLI-state?

Are any attributes shared by SLIders? One of our informants told us of receiving a severe electrical shock when he was a little boy: have other SLIders had such experiences?

Is the effect cumulative? For example, suppose two SLIders were to walk down a light lit street together, not concentrating but chatting idly of this and that, would they have a noticeably greater effect on the lights than one alone?

How widespread is the ability to affect other appliances? Every SLIder should be tested for any effect he may have on radios, TV and other appliances.

Female SLIders should be tested at different phases of their menstrual cycle.

SLIders should be questioned under hypnosis, to ascertain whether they are subconsciously aware of what they are doing.

Anthropologists should be invited to consider whether the SLI-force and machine-busting force bear any resemblance to the abilities of shamans and medicine men in other cultures. It would be interesting to know if such persons do SLI when confronted with streetlights.

Psychologists should draw up a questionnaire to determine whether SLIders share any common behavioral traits with one another, and by comparison with a control group of non-SLIders. What significance, if any, attaches to the fact that SLIders can be in so many different – and often contrasting – states of mind?

Parapsychologists should be consulted regarding the similarities – and the differences – between SLIders on the one hand, and PK and poltergeist subjects on the other.

Trying to make sense

Turning a light on or off is such a matter-of-fact action: we all do it many times in the course of every day. But the way SLIders do it, it isn't matter-of-fact; it is extremely odd. So how do we make sense of something so odd?

Most of our inquiry has been frustrating. We have dug into questions and found that, like Russian dolls, they have further questions hidden inside them. For example, we found that most SLIders can't do it when they *try* to do it – but occasionally, some can. Why the exceptions?

Nevertheless, we have established some near-certainties which, though they fall sadly short of resolving the SLI enigma as a whole and though they are not as rock-solid as we would wish, do at least seem to give us some firm ground on which to stand while we consider further implications. Let's try to summarize what we have learned.

Firstly, it is evident that SLI is not entirely arbitrary. It happens only to certain people, but it frequently happens to those people time and time again. We have yet to learn how and why the lights distinguish between the people who come near them, going out for some and not for others.

Secondly, it happens to them only when they are in a suitable frame of mind, and for the most part only when they are not thinking about SLI or trying to make it happen. We have yet to learn how and why the lights are able to discriminate in this way.

Vague as they are and meaningless as they seem, these distinctions appear to be the rules of the game. And the fact that rules exist at all tells us that who/whatever is making SLI happen is doing so with some kind of *intention* and perhaps of *purpose.*

But *whose* intention? Is there something in the light that responds to SLIders? A streetlight has no feelings: it neither loves SLIders nor hates them. It is incapable of intention or any kind of purpose, other than doing the job it was designed to do and resisting any attempt to prevent it – until a SLIder comes walking down the street. Whereupon it goes briefly berserk.

So no, we can rule the light out. It has to be the SLIder herself. Yet scores of SLIders have had this kind of experience:

> *Peewaukee, Wisconsin:* This week, driving with my girlfriend, I said I wanted to see if I could put out a light by trying. I kept concentrating but none would go out. However, as soon as I gave up (released the thought) one went out. [36]

Necessarily, then, the intention comes from some part of the SLIder. But since it is not consciously controlled by him, and since, indeed, it often acts *contrary* to his will, this leaves us with no other choice than to accept the existence of some other element in the SLIder's make-up, a "second self" operating subconsciously, independently of the conscious self, yet purposefully and with intention.

Intention? But SLI appears to be nonsensical: without meaning, without purpose, without making any contribution to the welfare, the enlightenment, or even the entertainment either of the individual or of mankind in general. What conceivable intention can it have?

Force field theories

We can be reasonably sure that SLI is not a vague force floating about that people just happen to bump into; it's clearly directed at us, or at some of us at least. Plentiful theorizing has played with ideas about the electrical force field that surrounds the human body and could, it is argued, play a crucial part in our interactions with the world around us. German-born sociologist Kurt Lewin (1890-1947) evolved a holistic field theory in which he supposed that we each carry about with us, as it were, a field comprising our motives, values, needs, moods, goals, anxieties, and ideals. Such a delightful catchall concept can easily be used to account for SLI, which would be interpreted as an outward expression of the individual's tendencies and attributes and his current state of mind.

As an account of SLI, philosophically considered, such a suggestion has obvious appeal. It gives us a plausible framework into which we can fit our speculations. But that's all it does. When we come down to the practicalities of SLI, these umbrella speculations are too up-in-the-clouds to help us. They don't tell us why streetlights are such a favorite target. They don't explain the variations in pattern, or no pattern at all. While they could conceivably tell us why some SLIders do it when they are stressed, they don't help at all with the diversity of response – why some people make cars blink their lights, or interfere with checkouts and crossing lights.

Precursors of the future?

When we first raised the question of why some people do SLI and others don't, we tentatively touched on the possibility that there might be a deeper, hidden purpose for this. It was becoming evident that the crucial factor in SLI is the individual who does it, and that the question of what it is done to is only a secondary part of the enigma. So we have to ask, again, what is so special about SLIders? To all outward appearance, they are just plain ordinary folk picked at random from the phonebook. They seem to stand out from the crowd in only this one way – that they do SLI.

Since there's no way we can place each SLIder on the psychiatrist's couch and set about probing the hidden recesses of his mind, let's try another approach. Suppose we give a thought to the people – most of us, it seems – who *don't* do SLI? What's so wrong with us that we can't join in the fun? Are we deficient in some respect? Are SLIders ahead of the rest of us on the evolutionary ladder? Could they be precursors of the future? Even though we ourselves may never become SLIders, will our great-grandchildren develop SLI-power as a matter of course, as part of their genetic inheritance? If so, we are the first generation of humans to be not only passing through an evolutionary stage but to be *aware* of it happening around us.

Such questions have been raised before, though in a different connection. Throughout human history, indeed, there have been people who looked for some deeper underlying cause for the enigmas of life. The result, usually, was the elaboration of some religious or quasi-religious belief-system that could encompass all kinds of wonder-working entities from gods and spirits of the dead to abstractions such as Primal Causes and Natural Law. Which was all very good as mind-stretching mental exercise but didn't present much in the way of practical guidance. Anything we couldn't understand was simply one of Mary Rose Barrington's "Just One of Those Things" on a broader scale.

In 1901 Canadian psychiatrist Richard Bucke published *Cosmic Consciousness*[38] in which he pictured the development of man as a three-stage process. The first stage is the simple consciousness of animals. The second stage is the self-consciousness of the mass of humanity, including the growth of reason and the faculty of imagination, which is where most of us stand today. We're pretty bright, some of us, but apt to be baffled when confronted with stuff like SLI.

The third stage in human development, Bucke suggests, is the process of what he terms "cosmic consciousness." This is the stage in which mankind learns to think more broadly about our role in the ultimate scheme of things. This process has already started, he proposes, and can be seen in progress here and there. Individuals who have in his view discerned or had intimations of this third stage are cited in his enormously influential book. His three-stage scheme is a bit too slick to be altogether convincing, but it is a reasonable way of describing what may indeed be happening. Throughout the 20th century, philosophers, scientists, and other thinkers formulated their own accounts of this development in man.

All this has been very fine in principle, and many thoughtful ideas and systems of ideas have been projected. Think Gurdjieff, Huxley, Teilhard de Chardin, Sartre, and many others. But there was little substance to bear them out. These schemes were all very well, but they seemed too theoretical, too detached from the puzzles and problems of everyday life.

As far as I know, UFOs and SLI have nothing in common, but both have inspired thoughtful minds to grapple with seemingly insoluble problems. Ironically, since UFOs have such a tenuous hold on reality, it was they who inspired some of the most interesting ideas about how cosmic consciousness works in practice. The literature of ufology is enormous; the apparent mystery has attracted speculation of all sorts, some of it downright silly, but much of it thoughtful and valuable. Two writers in particular, Jacques Vallee[39] and Leo Sprinkle,[40] have come up with speculative ideas that are relevant to our investigation. Both looked to some kind of cosmic purpose behind the supposed phenomenon. Vallee, more than anyone, taught us to recognize that UFOs are not nuts-and bolts space vehicles ferrying aliens from Alpha Centauri but may be better regarded as messengers from other planes of reality. His specific ideas are not relevant to our present inquiry; what is important was that he opened our minds to broader possibilities. Sprinkle, a professor of psychology at the University of Wyoming at Laramie, was more specific when he proposed that the purpose of the UFO phenomenon was to raise mankind to a higher level of awareness, whereby we would become "cosmic citizens." Whether their ideas are right or wrong, who cares? It's not the thought but the *way of thinking* that is important.

It is by no means ridiculous to wonder whether something of the sort may be going on with SLI – that, trivial as it may appear, it could be a part of the next development in man. If you're anything like me, you're put off by woolly speculations that have too much about them of the occult and the paranormal, when, for heaven's sake, all we want to know is what's happening to our streetlights. But the fact is that what's happening to them is inexplicable – it defies explanation to the point of being impossible. Faced with such a challenge, we would be daft to refuse to grab any lifeline anyone throws out to us, however flimsy it may look.

Theories of what cosmic consciousness might entail vary widely, from ancient Buddhist traditions to current developments in quantum theory. For our purposes, the suggestion is that by such means as UFOs and SLI, the mind is being "stretched" to entertain ideas that at first seem to lack any kind of feasibility. It is now widely accepted that the reach of the mind extends farther than is commonly thought, that it is active in ways which transcend the conventional limitations set by consensus science. There is no need to discard science, but simply to extend the boundaries we have set to it. To recognize that there are forces we don't yet comprehend doesn't require us to plunge into metaphysics, the paranormal, or the psychic, less still the occult or the magical. We have seen that SLI is not the only phenomenon of its kind; the poltergeist, especially, confronts us with much the same kind of challenge. If we simply set it aside as "just one of those things" we are doing science, and ourselves, a disservice.

We will all benefit if we find out what is going on in SLI. It will tell us more about ourselves, and in doing so will tell us more about the universe we live in. For, like it or not, our universe is one where things like SLI can happen, and *do* happen. To fail to understand what is happening in SLI is to leave a piece of unknown territory unexplored – and it could be an area of experience which it would be very much to our advantage to understand, revealing powers of the mind that could be beneficial, even life-transforming.

SLI is a relatively simple phenomenon, and investigation should, in principle, be relatively simple also. We know *what* happens in SLI. We know quite a lot about *how* it happens. But the really interesting questions are *why* does it happen, and *why* to some people and not to others, and *why* only when they aren't trying to do it?

Could SLI be trying to tell us "Look, you have inside you the power to do more, much more, than you think you can"?

If we will but put our minds to it...

ENDNOTES

1. Letter, 16 November 2000, in author's files
2. *Omaha World Herald*, January 1990
3. Richard Wiseman, personal correspondence, 28 August 1998
4. Auerbach, Loyd, 1986, *ESP, hauntings and poltergeists*, NY Warner: 32
5. Whitman, Walt, 1859, *Leaves of grass*, many editions
6. Monro, Jean, paper "Electromagnetic effects on man," (circa 1998) in the author's possession
7. Rhine, Joseph Banks, 1938, *New frontiers of the mind*, London: Faber & Faber; and other writings
8. Freud, Sigmund, 1901, *The psychopathology of everyday life*, Collected works vol vi. London: Hogarth Press
9. Fort, Charles, 1919, *The book of the damned*, New York: Boni & Liveright
10. Corliss, William R., 1982, *Lightning, auroras, nocturnal lights and related luminous phenomena*, Glen Arm: Sourcebook Project; Gaddis, Vincent, 1967, *Mysterious fires & lights*, New York: McKay; Rogo, D. Scott, 1990, *The poltergeist experience*, New York: Penguin, and many others
11. Report on investigation by Hilary Evans for Bristol UFO organisation
12. Angélique: Louis Figuier, *Les mystères de la science*, vol 2; Smyrna girls: Rogers, 1853, *Philosophy of mysterious agents*, Boston: Jewett; Rochas *Forces non definies*, 82 Mozzani, *Magie et superstitions*, 331+ tells of Cottin and gives other good cases
13. Among notable recent investigators of PK are Stephen Braude
14. "A spirit of anger" entry in *The unexplained*, partwork publication, London: republished 1992 as *Mysteries of mind, space and time*, Westport CT: Stuttman
15. BCE poltergeist. See Gauld, Alan & Cornell, Tony, 1979, *Poltergeists*, London: Routledge
16. Perrault, François, 1553, *L'antidémon de Mascon*, reprinted

1853 private

17. Rogo, D. Scott, 1990, *The poltergeist experience*, New York: Penguin
18. Laubscher, BJF, 1972, *Where mystery dwells*, Cambridge: James Clarke
19. Moor, Major Edward, 1841, *Bealings Bells*, Woodbridge: private
20. Pauli Effect: online Wikipedia
21. Playfair, Guy Lyon, 1980, letter in *Fortean Times* 31:54 containing translation of original account
22. Personal letter in Hilary Evans files
23. Part, Shepley, 1899, "Occultism in West Africa" in *Proceedings of the Society for Psychical Research*, London
24. MacManus, Daniel, 1959, *The middle kingdom*, London: Parrish
25. Fuller, Curtis, May 1961, editorial in *Fate*
26. Ward, Franklin D., May 1961, Letter in *Fate*, marsh gas letter
27. Cited in Cade & David, 1969, *The taming of the thunderbolts*, London: Abelard-Schuman
28. Rutledge, Harley D., 1981, *Project identification*, New York: Prentice-Hall
29. *Fortean Times*, 42 Autumn 1984
30. Entry "Aureole" in Addis & Arnold, 1916, *Catholic Dictionary*, London: Virtue
31. Gorres, 1834, *La mystique divine, naturelle et diabolique*, Paris: Poussielgue tome 2: 74-5 (translated from German)
32. For a striking depiction of this, see the Italian periodical *Tribuna Illustrata*, April 1934, and can be seen at maryevans. com, item: 10018230
33. Rex G. Stanford, "98" paper in *JASPR* 75
34. Jennings, Paul, 1950, *Oddly enough*, London: Reinhardt & Evans
35. Evans-Pritchard, F.E., 1937, *Witchcraft, oracles and magic among the Azamde*, Oxford University Press
36. Hamilton, T. Glen, 1942, *Intention and survival*, Toronto:

Macmillan
37. Housman, A.E., 1922, *Last poems.* London: Constable
38. Bucke, Richard Maurice, 1901, *Cosmic consciousness,* Philadelphia: Innes
39. Vallee, Jacques, 1992, *Forbidden science,* Berkeley CA: North Atlantic, and other writings
40. Sprinkle, R. Leo, 1999, *Soul samples,* Columbus NC: Granite

APPENDIX: SLIDERS

This list includes most of the SLIders who have told us about their experiences. Names and addresses are incomplete to protect their privacy, but it is interesting to learn where SLIders come from (pretty well everywhere!) and their occupation. Email correspondents often give little information. Considering how few people have had the opportunity to know of our investigation, the coverage is very wide.

Reference# | Name	Location (U.S. unless noted)	Occupation
001 Susan E	Hickory NC	housewife
002 David L	Eugene OR	
003 Steven Y	Fullerton CA	
004 Dr. George E	Budapest Hungary	engineer
005 Don I	Taylor MI	
006 Mary A	Minneapolis MN	singer
007 Eric W	Fair Oaks CA	TV art director
008 John S	Harpenden UK	writer
009 Renee W	Woodland Hills CA	nursing student
010 as Meketa	Plano TX	
011 Ronald B	Woodinville WA	chemistry engineer
012 H. J. B	Cheltenham UK	draughtsman
013 Yvonne B	Monument CO	mathematician
014 Lester L	Corona CA	
015 Martin C	Gainesville GA	
016 Dean S	Oxnard CA	
017 Anne de V	Central City CO	
018 Joan C	Round Rock TX	
019 Phillip S	Alexandria VA	computer engineer
020 Gwen M	Brooklyn NY	
021 Michael F	San Jose CA	electrician
022 David K	Lethbridge, Alberta, Canada	
023 Mary Ann M	Hercules CA	editor

024 Bill S	New York NY	alarm engineer
025 Richard F	Reno NV	geologist and taxi driver
026 Clayton J	Normal IL	
027 Teri M	Carlsbad CA	
028 Joseph L	Choektowaga NY	
029 Erin M	San Francisco CA	
030 Dennis M	Worcester MA	
031 Jon F	Yonkers NY	
032 Leonard M	Huntington Park CA	automobile technician
033 Craig P	Evergreen CO	musician
034 Herbe S	Los Angeles CA	
035 Bruce W	Rohnert Park CA	
036 Ed P	Pewaukee WI	
037 Daniel B	London UK	psychologist
038 Robert L	Great Falls MT	Air Force
039 David G	Kenmore	NY
040 Bernard L	Paris, France	teacher
041 Steve W	Augusta GA	
042 Monahan	Toledo OH	
043 John O	Elmira NY	artist
044 Karen P	Statesville NC	
045 Dennis B	Fulham, London UK	hospital worker
046 Candace K	Valencia CA	
047 Karl K	Fort Myers FL	minister, counselor
048 Graham R	Norwich UK	
049 G. T. S	London UK	
050 Raymond S	Marina Del Rey CA	
051 Vicki B	Spring TX	psychic medium
052 Edward F	Bricktown NJ	
053 Vickie B	North Vernon IN	
054 Naomi R	Pardeeville WI	radio announcer
055 Greg S	Santa Rosa CA	software engineer
056 Susan G	Ottawa ONT, Canada	registered nurse
057 Charles D	Claremore OK	chiropractor
058 Josette F	Hawaii HI	housewife

059 Ed H	Represa CA	student
060 Robert S	Fredericksburg TX	
061 Annette J-O	Athens, Greece	healer, clairvoyant
062 Karen R-S	Strood UK	educator
063 Graydon B	New Brighton MN	engineer medical equipment
064 Kathleen B	Madison WI	
065 Michael B	Honolulu HI	investigator
066 Allen B	Tempe AZ	attorney at law
067 Richard C	Los Angeles CA	
068 Marilyn D	Grandview Heights OH	
069 Michael D	Hayward CA	
070 Loreen F	Emerson NJ	housewife
071 Wolfgang H	Durnsen Switzerland	
072 Debra H	Arcata CA	
073 Glenn M	Falls Church VA	communications manager
074 John & Mary S	Priest River ID	
075 Edward B	Seaside CA	attorney at law
076 Sandra Yeagin	Woodville TX	
077 Moore [unknown]		
078 Timothy H	Portland NY	
079 Michael G	Windsor ONT, Canada	
080 G. J. H. (& wife Gloria)	Severn MD	
081 Geraldine O	Woking UK	electrical design engineer
082 Margie H	Birmingham UK	
083 Wade N	Petawawa, ONT, Canada	
084 David M	Leighton Buzzard UK	
085 Jose V	Pontevedra, Spain	teacher
086 Trudy T	East Wenatchee WA	artist musician
087 Stephen L	San Francisco CA	
088 Frank H	Dumont NJ	
089 Allen M	Harrison AR	staff sergeant, air force
090 Mark B	Jamesville NY	psychotherapist counselor
091 Margaret M	Newton Abbot UK	lawyer
092 Liam R	Worcester UK	engineering student
093 John G	Calera AL	

094 Ina W	Leipzig, Germany	electro engineering merchant
095 Heather B	Halifax, Nova Scotia, Canada	
096 des Moreno	Barcelonetta, Puerto Rico	hospital worker
097 Sergio L	Zaragoza, Spain	student
098 José M	Maren de Marguan, Spain	functionary
099 Juan G	Zaragoza, Spain	administrator
100 Howard B	Willowdale ONT, Canada	dermatologist
101 J. W. B	Chiswick, London UK	
102 D. L	Barriem ONT, Canada	
103 Keith B	Arroyo Grande CA	
104 Dolores W	Edmond OK	
105 I. B	Manchester UK	
106 Derek A	London UK	
107 A. C. B	London UK	
108 Sarah B	Croydon, Victoria, Australia	
109 Clive B	Aylesbury UK	
110 Stella B	Balham, London UK	
111 Robert C	Milton Keynes UK	
112 Chris C	Belgrave, Victoria, Australia	factory staff
113 Julie C	Edinburgh, Scotland UK	
114 Leah D	Green Bay WI	
115 Andrew D	Pontefract UK	law student
116 Gillian D	Stavanger, Norway	
117 Gareth D	Ipswich UK	
118 Jane de la R	Horsell Village, UK	
119 Finn	Hampstead, London UK	
120 Anthony G	Richmond, London UK	
121 David G	Melbourne, Australia	film/TV
122 Robert G	Hurstbridge, Victoria, Australia	
123 Christopher G	Birmingham UK	Anglican minister
124 Philip H	Belfast, Northern Ireland UK	
125 S. R. G. H	Walthamstow, London UK	
126 G Hildebrand W	Victoria, Australia	musician
127 Veronica H	Glen Iris, Victoria, Australia	
128 John H	Birmingham UK	

129 Karen K	Oxshott UK	
130 Bill L	Northampton MA	
131 Daphne M	Greenwich, London UK	
132 Richard M	London UK	
133 Nina M	Antony, France	
134 Sarah M	Hastings UK	
135 Peter M	Great Yarmouth UK	
136 Barbara M	Coulsdon UK	
137 J. Heidi N	Columbus OH	
138 Donna N-S	North Cadbury UK	management consultant
139 Tom R	Liverpool UK	
140 Colin R	Darlington UK	computer graphics
141 Helen S	Montmorency, Victoria, Australia	
142 Chris S	Anerley, London UK	electrical contractor
143 Heather S	Kingsbridge UK	
144 Andrew W	London UK	
145 Timothy K	Loxahatchee FL	
146 David A	University of Hertfordshire UK	astronomer
147 James G	Hampstead, London UK	student
148 Rose M	Stoke Newington, London UK	microbiologist
149 Margaret P	Cambridge UK	
150 Richard R	address unknown	
151 Rich T-N	Glen Cove NY	
152 Diana B	Copperas Cove TX	
153 Andrew D	Highbury, London UK	
154 Peter V	London UK	mathematician
155 Kathie K (& family)	Tustin CA	
156 Kirsty S	Carlton, Victoria, Australia	clairvoyant
157 Ellen R	Austin TX	
158 Clive E	Bournemouth UK	
159 Anonymous		
160 Morgan	address unknown	
161 Tiffany R	Carbondale IL	
162 F. de J	Den Haag, Nederland	
163 Jeff F	St Charles IL	self-employed sales

164 Max W	Putney, London UK	
165 Laurence D	London ONT, Canada	
166 Rebecca H	Scandinavia WI	publisher
167 C. J. M	Wimbledon, London UK	
168 Morag H	Isle of Skye, Scotland UK	
169 Deb W	Brighton UK	artist
170 Nicky A	Leeds UK	
171 Bernard D	Hackney, London UK	
172 A. W. H	Raynes Park, London UK	
173 Cynthia D	Plymouth UK	
174 Helen D	Manchester UK	
175 K G P	Exminster UK	
176 Alex T	Dalston, London UK	
177 E. I. G	Birmingham UK	telesales
178 Dan C	Harrogate UK	
179 Michael T	Preston UK	
180 R N B	Tacoma WA	
181 John S	Baker City OR	student
182 Anne B	Kawarren, Australia	
183 Staci N	Claremont CA	
184 Danny D	Fukushima-ken, Japan	
185 Richard M	Riverside CA	
186 Phillip M	Burlington ONT, Canada	
187 Kevin F	HMS Nelson, Portsmouth UK	
188 Louisana C	Beckenham UK	
189 Andy P	Howden-le-Wear, UK	
190 Richard B-W	Hampton VA	
191 McQuirk	address unknown	
192 Patricia A	Edinburgh, Scotland UK	
193 Gareth H	Norwich UK	
194 David T	Stourbridge UK	
195 James anon	Eire	
196 Randall S	Denver CO	
197 Susaidh M	Scotland UK	
198 Nick S	Aston Clinton UK	

199	Mark S	UK	
200	Annette G	CT	
201	Grainne M	UK	
202	Tracey S	Alexandria VA	
203	Kane B	Wichita KS	student at the time
204	Jim G	Gavle, Sweden	
205	Ruth anon	Canada	housewife
206	Anonymous	Somerset UK	
207	Jay S	Vancouver, Canada	
208	Jeff B	Phoenix AZ, regarding his wife	
209	Eric C	Mariakerke, Belgium	
210	Laura G	Washington University WA	(then) student
211	Anonymous	Saint Paul MN	teacher
212	Trevor O	London UK	
213	Simon S	UK	
214	Jorgen S	Norway	
215	Sergio Q	US	product engineer

CPSIA information can be obtained
at www.ICGtesting.com
Printed in the USA
LVHW111433240520
656468LV00001B/30